A Guide to...
Neophema & Psephotus
Grass Parrots
Their Mutations, Care and Breeding
Revised Edition

By Toby Martin

Published and Edited by Australian Birdkeeper Publications ©

© 1997 Australian Birdkeeper Publications

All rights reserved. No part of this publication may be reproduced, stored in any retrieval system, or transmitted in any form or by any means without the prior permission in writing of the publisher.

**First Published 1989 by
Australian Birdkeeper Publications
PO Box 6288
South Tweed Heads
NSW 2486 Australia
Reprinted 1990 and 1992
Revised Edition 1997**

ISBN 0 9587102 44

Front Cover:
Top left: Opaline Red-rumped Parrot - A. Chalmers
Centre left: Naretha Blue-bonnet - B. Branston
Bottom left: Pink Bourke's Parrot - B. Branston
Bottom right: Pied Hooded Parrot - B. Mullens
Back Cover:
Lutino Scarlet-chested Parrot - T. Martin

Design, Type and Art: PrintHouse Multimedia Graphics (Gold Coast)
Colour Separations: NuScan (Gold Coast)
Printing: HBM Print (Brisbane)

Contents

About the Author .... Page 5
   - Personal Note from the Author 5
   - Acknowledgements 5
About this Book 6

NEOPHEMA GRASS PARROTS

Introduction 8
Management
   - Housing 8
   - Feeding 11
   - Breeding 13
   - Nestboxes 14
   - Hybrids 15
   - Ringing 15
   - Keeping Records 16
   - Surgical Sexing 16
TURQUOISINE PARROT 17
   - Sexing
   - Mutations
SCARLET-CHESTED (Splendid) PARROT 24
   - Sexing
   - Mutations
ELEGANT PARROT 32
   - Sexing
   - Mutations
BLUE-WINGED PARROT 35
   - Sexing
   - Distinguishing the Elegant Parrot from the Blue-winged Parrot
   - Mutations
ROCK PARROT 36
   - Sexing
   - Breeding
   - Mutations
ORANGE-BELLIED PARROT 37
BOURKE'S PARROT 38
   - Sexing
   - Mutations

PSEPHOTUS GRASS PARROTS

Introduction 44

Management
- Housing 44
- Feeding 45
- Breeding 46

RED-RUMPED PARROT 46
- Feeding
- Breeding
- Mutations

MULGA PARROT 57
- Breeding
- Sexing
- Mutations

BLUE-BONNET PARROT 60
- Housing
- Feeding
- Breeding
- Mutations

HOODED PARROT 62
- Housing
- Feeding
- Breeding
- Mutations

GOLDEN-SHOULDERED PARROT 67
- Breeding
- Mutation

PARADISE PARROT 68

DISEASES OF NEOPHEMA AND PSEPHOTUS GRASS PARROTS

Introduction 70
- The Health Check of Newly Acquired Birds 70
- Quarantine of New Birds 70
- Stress and Disease 71
- Recognising Disease 71
- Control and Prevention of Disease 71
- Prevention of Disease within the Aviary 72
- Worm Treatment 72
- Worming Programme 72
- Dose Rates for Deworming each Species 74

DISCLAIMER 74
PUBLISHERS' NOTE 75
DOUBLE HOSPITAL CAGE AND BROODING BOX 75
HANDREARING 77
- Recipes

GENETIC TABLES 79

About the Author

Toby Martin was born into a family of bird breeders in Sydney, Australia in 1927. His early years were spent working as a shipwright on the Sydney waterfront and travelling the world as what could best be described as an adventurer, dealing in shells and all sorts of rare and unlikely objects. Toby is also quite an artist, specialising in painting the older sailing ships.

He married in 1950 and after 22 years his wife found his passion for birds too much to bear. Some years after his divorce, he met his present wife, Jacki, who also bred birds. It was with her that he really specialised in the breeding of Neophema and Psephotus Grass Parrots. He now keeps many of the mutations available today and of course the colours that will combine to produce the mutations of the future.

Toby lectures at many avicultural meetings all over the world. He has been invited to lecture in England, Sweden, Denmark and the United States of America. He also founded, and is the current President of The Grass Parrot and Lorikeet Society of Australia Inc.

A Personal Note from the Author

I would just like to explain that I suffer from dyslexia and as such, have great difficulty reading and writing. This book is an example of what can be achieved through determination and the assistance of friends. There are many experienced aviculturists who have a wealth of knowledge to share but for one reason or another are unable to put it down on paper. For the betterment of aviculture I would strongly urge such people to contact **Australian Birdkeeper Publications** who will do all they can to assist, as they have done for me.

Acknowledgments

The author and publisher would like to take this opportunity to thank the following people for their assistance in the preparation of this book, Ian Brown, Bill Boyd, Walter Boles, Curator of the Ornithological Section of the Australian Museum, Sydney, Alan Chalmers, Lee Ford, Graham Matthews, Bob and Betty Branston, Peter Rankine, Colin O'Hara, Bruce Brown, Colin Sylvester, Bruce Mullen, Joe Saliba, Peter Ronnberg, Gordon Dosser, Nick Livanos and Peter Brown for his information on the endangered Orange-bellied Parrot.

We would also like to thank Dr Terry Martin for his assistance in regards to the genetic inheritance of mutations and the overview of the Diseases chapter in this book.

Special Acknowledgment from the Publisher

Together with his wife, Jacki, Toby Martin displays the true passion of a genuine love and dedication to birds - a true aviculturist. It is with great honour and appreciation of a special aviculturist that I have had the opportunity to publish this revised edition.

Toby, thank you for sharing your vast knowledge with bird keepers worldwide.

About this Book

This book is full of information to allow you to get as much pleasure from these birds as possible. Nobody likes losing their birds, everybody enjoys breeding them. Unless the birds are happy, both in their environment and in themselves, they will not breed successfully. The details you can glean from this book should be put into practice. Some vital points will be repeated throughout the text.

This book has been written for aviculturists, therefore it concerns birds that have been bred in aviaries and consequently details of their distribution or activities in the wild are minimal.

Every method of breeding these birds are by no means covered by this book. However, the methods discussed have all been very successful. A successful method must be more beneficial to the reader than numerous possible alternatives.

Aviary-bred parrots should be viewed differently to wild parrots, in the same way as the aviary Budgerigar and domestic chicken are to their wild counterparts. Only aviary-bred birds should be considered for aviaries. Having said that, such birds are our responsibility and it is up to us to learn how to provide the best for these lives, dependent on our care.

This book has tried to avoid the ambiguous, over-generalised statements often seen in other publications. Read, enjoy and learn. Your birds are depending on you.

Within Australia all birds discussed in this book are protected under law, enforced by each state or territory National Parks and Wildlife Service. Their advice should be sought if you are unsure of the procedure to follow before acquiring these birds.

The revised edition of this book is essentially the same as the original with the addition of more colour images of the newer mutations. It also looks at the various types of aviaries and conditions that both Neophema and Psephotus Grass Parrots are kept and bred in worldwide.

The Neophema Grass Parrots

Cock Turquoisine Parrot

Introduction

These pretty little parrots have everything going for them. They are attractive, relatively quiet and peaceful, and can be mixed with finches and other non-aggressive parrots such as the Princess Parrot. Being smaller parrots, they do not require the building of large aviaries necessary for the larger parrot and cockatoo species. With a few exceptions they are usually willing to breed.

They have a very reasonable life span. I have a 17 year old cock Turquoisine Parrot and hen Bourke's Parrot that has been breeding for ten years.

Some aviculturists will be pleased to know that there are now quite a variety of Neophema mutations. Some are available now, others are still relatively rare at this moment in time. The mutations will be discussed under the relevant bird species.

They are definite family favourites - and gentlemen, these are often the birds that are the key to success when trying to persuade your wife to move the washing line a bit to the left in order to accommodate a few more aviaries.

The six members of this family are:

Turquoisine Parrot *Neophema pulchella*
Scarlet-chested (Splendid) Parrot *Neophema splendida*
Elegant Parrot *Neophema elegans*
Blue-winged Parrot *Neophema chrysostoma*
Rock Parrot *Neophema petrophila*
Orange-bellied Parrot *Neophema chrysogaster*

Since the first edition of this title was released in 1989 the Bourke's Parrot, after much research, has been reclassified to its own genus. Although we have retained the bird under the Neophema section, it is accepted that the **Bourke's Parrot** *Neopsephotus (Neophema) bourkii* is no longer classified as a member of this family.

Management

Housing

In a longish aviary some brush will give advanced warning of the end wire to young birds.

My aviaries for breeding single pairs of Neophema Grass Parrots measure 2.1 metres (7 feet) long by 1.8 metres (6 feet) high and 800mm (31.5 inches) wide. This allows me to get eight such aviaries in an area approximately 6 metres (19.7 feet) long. Originally my aviaries were 3.6 metres (12 feet) long but I have found Neophemas do well in the shorter flights.

These 2.1 metres (7 feet) aviaries not only suit the birds but also save me space and money. One of the main points in favour of shorter aviaries for the Neophema and Psephotus Grass Parrots is that both adult and young birds are capable of picking up speed very quickly, and in long aviaries fatal accidents often occur when contact is made against the often unseen front wire. This is especially the case in young birds who are unfamiliar with their new surroundings. If your aviaries are already quite long, this problem can be lessened greatly by placing soft branches of tea-tree or pine in front of the end wire.

Above: Rear feeding doors.
Below: Neophema Grass Parrot aviaries.

The shelters are 1.2 metres (4 feet) long by 800mm (31.5 inches) wide and 1.8 metres (6 feet) high. The shelter is divided from the flight at ground level with 450mm (18 inches) high tin or steel. This prevents vermin entering the shelter. To stop rain blowing in, 20mm (8 inches) of tin drops down from the roof. The fibro ceiling is insulated with 75mm (3 inches) Insuwool™ below the galvanised roof. The concrete floors have 5mm (2 inches) of course river sand over them.

The fibro walls and ceiling inside the shelters are painted light green. I use Acrylic Industrial Flat White paint, tinted green, which is cheap, washable and harmless to the birds. The majority of paints available today do not contain lead. This brightens up the shelter and does not discourage birds from entering. Dark shelters, in comparison with the outside flight, are viewed with suspicion by many birds.

The shelters, after all, are where I want my birds to breed, feed and generally retreat if disturbed or if the weather is unfavourable - therefore they must be very acceptable to them.

The shelters back on to a 1.8 metre (6 feet) wide enclosed walkway. This area allows me to set up hospital and quarantine cages and to store general equipment against the opposite wall, conveniently. By entering the aviary from the rear, the birds never have to fly at you, so this causes them less or no stress at all. Seed hoppers are attached to the 900mm (3 feet) high back door. They have a wire tray that allows seed and husks to drop into a catching bin below. This makes for a cleaner aviary and also allows the good seed, normally wasted on the ground, to be reused. If this is done, as a disease prevention, the 'recycled' seed should only be used in the aviary from which it is collected. Next to the hopper a wire basket/shelf holds fruit, vegetables, sprouted seed and Arrowroot™ biscuits etc.

I use natural branches of varying thickness for my perches as this exercises the birds' feet and also allows them to choose the most comfortable width on which to roost.

There is a fixed perch in the shelter. In the outside flight, however, I use a swinging perch. The ends

Right: Rear door with feeding hopper.

1. Interior of Alan Chalmers' aviary complex.
2. Colin O'Hara's aviaries in winter (UK).
3. Colin O'Hara's aviaries in summer (UK).
4. Alan Chalmers' aviary complex.
5. Peter Ronnberg's aviaries in Sweden.

do not come into contact with the wire sides. This I find helps stop the cock from squabbling with the cock next door when he should be paying attention to his hen. In order to squabble they have to hang onto the side wire. They find this awkward and don't really like it. It also encourages them to exercise, especially when keeping balance on landing. Overweight birds often fail to breed, and this must be watched for.

The aviaries should face north to northeast. The solid back therefore faces the cold southerly winds and the front benefits from maximum sunshine. Draught and damp are enemies of our birds and they should be excluded from any aviary. A friend of mine once pointed out that if you have a nail hole or similar in the side of your shelter, wherever it is, a Scarlet-chested Parrot would find it and sit in the draught and get pneumonia - so be warned! I do have one bank of aviaries of the same size completely roofed and enclosed with fibro, apart from the top 900mm (3 feet) at the front. These aviaries are excellent for problem or newly acquired birds - especially those from a warmer climate. They are also a good idea if for some reason your

The rear of the author's aviaries showing security grills and high boundary fence

aviaries cannot face the ideal direction. Late or early breeders can use these aviaries when the weather tends to be inclement. These aviaries also allow argumentative birds to be housed alongside one another. Unable to see the bird next door, there is less fighting and more breeding. Quite often, because they can hear one another, cocks of the same species often encourage each other into breeding condition.

I also have holding aviaries for young and non breeding birds. These are 6.6 metres (21.6 feet) long, 900mm (3 feet) wide and 2.1 metres (7 feet) high. The shelter is 2.1 metres (7 feet) long x 900mm (3 feet) wide x 2.1 metres (7 feet) high. These aviaries will house up to 30 Neophema Parrots. Soft bushy branches are placed on the end wire giving the young birds plenty of warning to slow down.

It is worth mentioning at this point that the Neophema group are generally very hardy birds and can adapt to many varied climatic conditions. This is borne out by the many birds that are bred overseas, in particular Europe, England and the United States.

A word of warning is relevant here. One morning I went down to my young bird holding aviaries and found odd Bourke's, Turquoisine and Scarlet-chested Parrots lying dead on the floor. After inspection, I found they had all died from cerebral haemorrhage. I couldn't work it out. My first thoughts were that cats or owls disturbed them during the night. When I informed an old breeder friend of mine, he immediately asked if I was housing Elegant Parrots in the same aviary. Yes, I told him. That's your answer he said, 'Go down to your aviary at dusk and you will see the Elegants bombing the other birds and hitting them on the head with their beaks. This I did and saw it for myself. Since then I have never held Elegant Parrots in with other Neophema species.

Observation of your birds is vitally important as a responsible aviculturist. Never assume that because all your systems are automatic and that you have a low maintenance setup, that daily inspections of all your birds is not as important. Ultimately, bird keeping should be a pleasure for the keeper, rather than a chore. However, responsibility to your charges to ensure their complete well-being will make the necessary work that much easier and rewarding.

Feeding

Seed quality must always be a serious consideration. To avoid substandard seed, I suggest you first find a reliable seed source. Then before buying a full bag or bags of seed, buy one or two kilograms. Take it home and soak it for no more than 12 hours. Spread it out on a cloth and allow it to sprout. (Sprouting jars are available for this purpose.) Before I buy any more seed, 70% - 80% must have successfully sprouted. If only 50% of seed sprouts, then half the seed is dead. This is only wasting your money and the birds merely benefit from half of what is fed to them.

My basic seed mix consists of 6 parts canary, 1 part French white millet, 1 part grey-striped sunflower (I prefer the small sunflower myself) and 1/2 part of hulled oats. I buy seed only three or four times a year and store it in 40 gallon drums.

Above: Alfalfa, mung beans and grey-striped sunflower seed are soaked in separate containers.
Below: Seed hoppers and 'extras' tray are easily serviced when the door is open. Water bowl is just inside the door.

Above: Drinking water is provided in white porcelain microwave oven dishes.

In addition to dry seed, I feed a lot of sprouted seed. I have 40 aviaries and in the breeding season they take 5 - 8 litres (1.3 - 2.1 gallons) of sprouted seed per day. Grey-striped sunflower seed, alfalfa and mung beans are soaked in separate containers for about 12 hours (usually overnight) - never any longer. Day 1 - they are then rinsed and left in large sieves to sprout. Day 2 - they are rinsed again. Day 3 - the sprouts should be almost 12mm (1/2 inch) long. Once over 12mm (1/2 inch) the sprout is past its prime and becomes drier. It is again rinsed and stored in a fridge for about six hours prior to feeding.

The sprouts are then finally rinsed and soaked for 30 minutes in an anti-bacterial solution. All three sprouts are then mixed and placed in a 5 litre plastic bucket. To this mixture I stir in two heaped teaspoons of vitamin and mineral powder, Vetafarm™ or Ornithon™. The sprouted seed is then given to the birds between 3.00pm and 4.00pm. This I find is important - by feeding sprouted seed in the afternoon, any that is left over will sit overnight and be eaten first thing in the morning. In the heat of the day the sprouts will dry up and be wasted. The young also benefit from a good feed early in the morning. During this afternoon feeding, I also give the birds 1/4 to a full Arrowroot™ biscuit and whatever is available at the time in the way of apples, endives, broccoli and carrots, cut or shredded into manageable pieces. I also check the seed hoppers and attend to the shell grit and cuttlefish supply.

In the morning the birds are given seeding grasses, milk thistles and dandelions when available.

During the breeding season, parents raising young should be fed daily.

I use white porcelain microwave oven dishes approximately 300mm (12 inches) square and 50mm (2 inches) deep for drinking water as these are easily cleaned. They are placed on bricks on the ground, well away from perches, and fresh water is given daily.

I also use calcium blocks made from moulding plaster (similar to plaster of Paris) in the drinking water. Some birds do not eat cuttlebone and are therefore not absorbing enough calcium, especially during breeding season. Hens can suffer from egg-binding and brittle bones, as a result of depletion of their natural calcium due to egg laying. By using calcium blocks in the drinking water the birds are more likely to receive their calcium requirements.

Breeding

In New South Wales, Australia, the Neophema breeding season commences in early to mid July and continues through to the end of March/beginning of April. All Neophema Grass Parrots perform the same basic courtship display. The cock struts and hops around the hen, bobbing up and down, stretching up to his full height and drooping or raising his wings and fanning his tail whilst giving his twittering call. His head bobs up and down as he regurgitates food and feeds his hen. Neophemas will breed from 8 - 12 months of age, although I prefer hens to be at least 12 months old, and cocks seem to do better if at least 18 months old. Mating will take place on the perch or on the ground. In all Neophemas only the hen broods. I have had no real success breeding Neophemas in trios. One pair per aviary undoubtedly brings the best results. The only exceptions to this are Rock Parrots and Blue-winged Parrots, where breeding in colonies is much more successful.

The method of breeding I use to great effect is to pair birds and allow the hen to have two good clutches. If the clutches are small, I sometimes allow them three clutches. I do stop her if she has reared eight or more babies. I had a hen Bourke's Parrot rear 16 young in one season. The following season she only reared two, so don't take too much out of your hens or your following season may be disappointing.

I carry spare cocks and hens in holding aviaries. When I stop the original hen from breeding, I leave the cock in the aviary and give him another hen, and she is nearly always desperate to go to nest. I then take two clutches from her, remove the pair and start again with a new pair. The incubation period in Neophemas varies between 19 and 25 days. However, many factors can affect this incubation time, for example, cold weather, when the hen may not start to incubate properly until the third or fourth egg has been laid, or the first three eggs may be infertile etc. Realistically and to be safe, the time from when the first egg is laid to the last chick has hatched is anywhere between 18 and 30 days.

The young stay in the nest between four and six weeks. When they fledge, they should be left with their parents for about four weeks. They will obviously have to be removed earlier if being persecuted by the cock.

For the most part Neophemas are non-aggressive although you may come across the odd rogue bird. You will just have to watch out for this. Many people think their pairs are incompatible because the cock and hen squabble and bicker between themselves, but I have found these pairs to hatch and rear more young than the really peaceful or docile pairs do.

Quite often you may find that for two seasons in a row you will breed a majority of hens, then in the third season a majority of cocks. This phenomena also occurs in the wild and I'm afraid I have no explanation for this.

One opinion with which I do agree, concerns the theory that hens can control the fertility of their eggs. If in a breeding season the hen sees a lot of rain which she associates with plenty of food (like seeding grasses etc) she may lay six eggs of which four to six will be fertile. In a bad dry season she may not have one fertile egg. I have

A nest of Turquoisine Parrot chicks.

had numerous hens which have reared eight to ten young for two consecutive seasons. Then, in a dry year, the same hens laid about 20 eggs each and fertility ranged from zero to two. I am sure it is not the cock's fault as they mate normally. The hen just has this 'control' over egg fertility.

Most Neophemas generally breed well. Blue-winged Parrots are an exception, as one season can be good and the following two dismal. They do fluctuate a lot from year to year. The other exceptions would be some of the newer mutations. To clarify this a bit more I would go as far as to say that out of 25 pairs of Normal Neophemas a good season should produce between 120 and 140 young.

From 25 pairs of a well-established mutation, such as the Red-fronted or Yellow Turquoisine Parrot, a good season should produce 70 to 80 young.

From 25 pairs of the newer mutations, such as Rose, Pink or Cream Bourke's or Cinnamon Elegant Parrots, 40 to 50 young produced would indicate an excellent season.

From this it can be seen that none of the Neophema mutations are as strong as the Normal at present. (More will be discussed on mutations under the specific species chapters).

Nestboxes

At present I am using two types of nestbox. The reason for this is that some of my birds prefer open top style boxes and some prefer dark enclosed style boxes. Therefore, I give the birds a choice and then remove the one not being used. I will describe both boxes. The enclosed nestbox is 200mm (8 inches) square and 300mm (12 inches) high. The hole is centered about 60mm (2.4 inches) from the top and is 55mm (2.2 inches) in diameter. A piece of 12mm (0.5 inches) dowelling sits below the entrance hole protruding 75mm (3 inches) outside and 60mm (2.4 inches) inside the box. A length of 50mm x 6mm (2 inches x 0.2 inches) timber is wrapped in 12mm x 12mm (0.5 inches x 0.5 inches) wire and fixed to the inside front of the box. The hen and the young can easily climb up to and sit on the inside dowel to be fed by the cock through the entrance hole.

The two types of nestboxes used by the author for his Neophema Grass Parrots. The pyramid has no lid

In summer these boxes can sometimes get too hot. This can be alleviated by sliding the lid back or propping it up, leaving a 25mm (1 inch) gap to allow hot air to escape. In this condition, attention must be paid to the weather. If you get a cold snap in the late afternoon or at night, the chicks can be fatally chilled if the hen is out of the nestbox. I hang these boxes about 40mm (1.5 inches) from the roof.

The open top model of nestbox is again 200mm square (8 inches) and 300mm (12 inches) high, but the front and back tapers to a wedge shape at the top. There is no hole. Instead the top is

The timber and wire ladder fixed to the inside of nestbox

open, forming a 200mm x 75mm (8 inches x 3 inches) slot. The same wire-clad climbing timber ladder is used inside. No dowel is used as the birds sit on the top. This box is hung about 150mm (6 inches) from the roof, allowing easier access by the birds.

These boxes hold the warmth at the bottom in colder weather, and in summer, allow the hot air to escape through the open top. There are no chilling problems.

Both boxes are made of 18mm marine plywood. They can be cleaned and disinfected and will last for years. (Chipboard holds the dirt and gets damp easily. If used, they should be replaced each season).

I do not like hollow logs because they are difficult to inspect and impossible to clean properly. They are therefore a potential hazard to the health of your birds. They are also less convenient to use than nestboxes.

The bottom of my boxes are filled to a depth of 60-75mm (2.4-3 inches) with a mixture of 75% contents of rotten logs and stumps found in the bush and 25% Oregon sawdust. I dampen the mixture with a few drops of water. Neophemas like humidity. During the breeding season I often remove the eggs, add a few drops of water and work it into the mixture with my hand, then replace the eggs. This doesn't seem to worry the hens at all. In nesting hollows in the wild, the rotted material soaks up moisture from the ground and the hen is constantly turning the mixture over, which dries rapidly. I do not use peat moss as I find it too dry.

When I first put the boxes in the aviary I place a square wooden frame in the bottom of the box. This reduces the area from 200mm (8 inches) square to about 140mm (5.5 inches) square. As the chicks grow or if the clutch is large, or if in hot weather the chicks try to keep away from one another to keep cool, I can remove this frame and give them the extra space they may require.

Hybrids

Hybridising or the breeding together of two different species of Neophema must be avoided. To my knowledge, no hybrid bred from a Turquiosine and Scarlet-chested parent have proved fertile. Hybrids from Elegant and Blue-winged Parrots have occurred.

Ringing

Whenever I pair up two Neophemas I allocate a specific colour of plastic split ring for their young. I hang these rings above the door of their aviary. When the pairs are changed or the hen is changed, so is the colour of the rings. All these records are written down.

When the young are in holding cages and I want to catch an unrelated pair, all I have to do is catch a pair with different coloured rings. With one-coloured numbered rings, I might have to catch several birds before finding an unrelated pair. This causes unnecessary stress to many birds.

Elegant Parrots are rung with split metal rings because they nearly always chew the plastic rings off.

There are many full colour and striped rings available, so many combinations are possible.

If you only have two or three pairs of a particular species of Neophema and you can find the right coloured rings, a useful way of keeping track of 'families' of birds at a glance is to ring the cock with say a green ring, the hen with a white ring and all the young with a green and white striped ring.

Keeping Records

Written records must be kept when breeding birds. Records should inform you of related birds or what bird is split to what colour. In addition, recording dates, times, weather conditions, temperatures, nestbox sizes, ages of birds, diets, observations, number of eggs laid, number of eggs hatched, number of young reared each season etc, provides valuable research for both your birds and aviculture. This information can reduce or eliminate infertility, poor parents, weak youngsters and generally substandard birds within your collection and indeed assist aviculture overall. Also, answers to why one breeding season is better than another can only be hoped to be answered from correlating relevant information recorded by hundreds of breeders like yourself.

Surgical Sexing

Many of my birds are surgically sexed. This procedure not only ensures me definite pairs but also indicates any problems or deformities in their sex organs. From this inspection it is possible to identify potentially infertile birds. Surgical sexing is a great aid to the serious breeder and its use cannot be over emphasised.

In conclusion, I must stress that birds quickly adapt to a routine. Changes to this routine often upsets them, so this should be avoided wherever possible. Good management is the key to success. To rely on 'Lady Luck' is to court disaster.

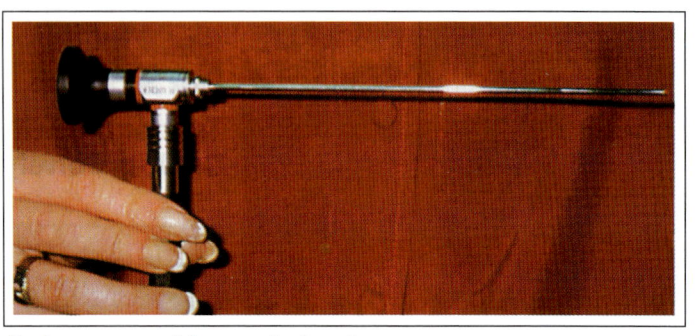

Endoscope used for surgical sexing.

TURQUOISINE PARROT
Neophema pulchella

The Turquoisine Parrot is classed as a free breeder and is well established in Australian and overseas aviaries.

This Neophema runs a close second to the Scarlet-chested (Splendid) Parrot for colour and beauty. It must be mentioned, however, that the Turquoisine is more aggressive than other Neophema Parrots. Adult pairs should be housed singularly. The only Neophema species that could be considered for sharing an aviary with a pair of Turquoisines is the Bourke's Parrot. It is preferable that the neighbouring flights do not house other Turquoisine or Scarlet-chested Parrots. If they must, then a solid opaque partition will avoid bickering through the wire. The swinging perches discussed under *Housing* also help when no such partition exists.

Up to six young in the nest is normal but the young should be moved when independent, in case they are persecuted by the cock, or sometimes the cock should be removed.

Sexing

The hen Turquoisine is generally paler than the cock bird. She has less and not such a rich blue around the face. The blue on her wings is less and paler than the cock. The most obvious difference, however, is that the hen lacks the red/chestnut blaze on the shoulder. (It should be said here that on rare occasions hens have appeared carrying this blaze.) As they colour up, young cocks are brighter than hens and as soon as a red feather is seen coming through on the shoulder it is virtually certain to be a cock.

Another problem often encountered is distinguishing the difference between Turquoisine and Scarlet-chested hens. In these cases, it should be found that generally the hen Scarlet-chested is more thick set, has more blue on the face, the lores are blue (the lores of the Turquoisine hen are whiter) and the hen Scarlet-chested has skyblue wing coverts whereas the hen Turquoisine wing coverts are a darker blue.

Mutations

Nearly all aviculturists are familiar with the Normal and Red-fronted Turquoisine Parrots. There are now a few more varieties being produced in Australian aviaries.

Full Red-fronted

This is a lovely bird. The red stretches from under the chin right down to the lower abdomen. The green and blue colourings also appear stronger and brighter than the Normal. The hens only show red half way up their bellies, not on their chests. For those who like Scarlet-chesteds but find their aviaries or climate does not suit them, this bird may make an admirable substitute. This mutation is dominant over the yellow belly of the Normal bird but requires selection to improve the red colour.

Normal cock Turquoisine Parrot.

1. Cock Jade Turquoisine (Red-bellied).
2. Yellow Pied Turquoisine.
3. Yellow Turquoisine Parrot cock.
4. Isabel Turquoisine cock and Yellow Turquoisine cock.

Yellow
This striking bird will only become increasingly popular in our aviaries. It is a recessive mutation that is well established.

Full Red-fronted Yellow
This is a combination of the Full Red-fronted and the Yellow Turquoisine. Red feathering on yellow tends to make the red look more orange. The strength of red on the front of this mutation differs from bird to bird and also in different shades of light. Pairs should be selected for breeding by the strength of red markings they possess. The stronger the red, the better.

Isabel
This mutation is still fairly rare. They have red eyes and are recessive. When established, they will be combined with other mutations.

Jade and Olive
Like green Budgerigars and green Peachfaced Lovebirds we now have three shades of green in the Turquoisine. The Normal has no dark factors, the Jade has one dark factor and the Olive has two dark factors.

The dark factor has been combined with yellow to produce new and interesting shades. It makes the yellow deeper and the blue richer in the Jade Yellow and light grey in the Olive Yellow. This mutation will be of further use when Blue and Pied Turquoisines appear.

Pied
These beautiful birds are still rare at the time of writing. They are a recessive mutation.

Other Mutations
Turquoisines have been bred showing a yellow flash on their wings, rather than a red one. When established, these birds could be used to eliminate the red flash from the Yellow Turquoisine, producing a more yellow bird.

The Blue Turquoisine, as far as is known, has been occasionally bred overseas but there has been no success producing blues in second and third generations. There are reports about such birds here in Australia, however, they have not been proved to exist, at the time of writing.

Full Red-fronted Turquoisine cock.

Full Red-fronted Yellow Turquoisine hen.

Above: Full Red-fronted Turquoisine cock.
Right: Cock Jade Turquoisine Parrot.
Below: Beautiful Tangarine Red Yellow Turquoisine cock. Combination of Red-fronted Yellow and Opaline.

R. COLLYER

1. Red-bellied Yellow Turquoisine cock.
2. Opaline Turquoisine Parrot
   (formerly known as Pied in Europe).
3. Pied Turquoisines - cock (below) and his son.
4. Pied Turquoisine (heavily marked).

1. Olive Turquoisine Parrot.
2. Isabel Red-eyed Turquoisine cock.
3. Yellow Pied Turquoisine Parrot.
4. Turquoisine Parrot eggs.

1. Olive Turquoisine Parrot.
2. Full Red-fronted Yellow Turquoisine cock.
3. Turquoisine hen.
4. Turquosine chicks in nestbox.

Scarlet-chested cock.

SCARLET-CHESTED (SPLENDID) PARROT
Neophema splendida
Other name: Splendid Parrot.

This is undoubtedly the most colourful of the genus. Its beautiful plumage contributes to its worldwide popularity.

This little 200mm (8 inches) long gem has broken many beginners' hearts. Many aviculturists describe it as a 'soft' or weak bird. It must be emphasised that poor housing with inadequate protection from damp and draught has caused most of the problems for this bird. Scarlet-chested Parrots do not like humidity. Better results are obtained in aviaries where the days are hot and dry and the nights are cold and dry. If your birds are uncomfortable in open aviaries then they must be enclosed.

Sexing

The most obvious difference is that the hen lacks the scarlet chest of the cock. The hen also has less and duller blue colouring on the face.

In young birds the cocks are brighter in colour and any sign of a red feather on the chest will quickly identify them as cocks.

Mutations

Red-bellied
This is a recessive mutation where a red belly has been added to the cocks red chest producing a Full Red-fronted Scarlet-chested. The hens show only the red belly.

Sydney Blue
These birds are also called the Sea-greens. They have a green back, cream belly and salmon pink chest. They are recessive. Hens do not show the salmon pink chest.

Par Blue
The back is greeny blue, the belly is creamy white as is the chest. A hint of pink can sometimes be detected on the chest. These too are recessive. Hens do not show the cream chest.

White-fronted Blue
As its name suggests, the cock of this mutation is blue with a white chest and belly. It is the most attractive and desirable of the Blue Scarlet-chesteds. Like the other blue mutations, it is recessive. Hens do not show the white chest.

1. Lutino
   Scarlet-chested cock.
2. Pied
   Scarlet-chested cock.
3. Cinnamon
   Scarlet-chested cock.
4. Albino
   Scarlet-chested hen
   and White-fronted
   Blue Scarlet-chested
   cock.

1. Scarlet-chested hen.
2. Scarlet-chested Par Blue hen in nestbox.
3. Golden Yellow Scarlet-chested,
   Ivory Scarlet-chested,
   Cinnamon Scarlet-chested,
   Sea Green Blue Scarlet-chested in nest.
4. Full Red-fronted Lutino Scarlet-chested cock and Normal coloured cock.
5. Isabel Scarlet-chested hen.

Isabel

This rare bird is recessive and is now established in Australia. It is different to the overseas Isabel which is sex-linked and may be misnamed, although Isabel has never been defined.

Other Mutations

Olive Scarlet-chested Parrots are established in Australia and are dominant.

Other known mutations include Lutino, Albino, Cream and Black-eyed White.

Combinations from Overseas

Cinnamon Blue - incorrectly called Silver overseas.
Skyblue   - genetically Isabel Blue.
Golden Yellow   - genetically European Isabel Cinnamon.
Ivory   - genetically Isabel Cinnamon Blue.

1. Australian Pied Scarlet-chested.
2. Blue Isabel Scarlet-chested Parrot.
3. Isabel Scarlet-chested cock.
4. Lutino Scarlet-chested hen and cock with Normal hen.
5. Skyblue Scarlet-chested Parrot.

Page 27

1. Golden Yellow
   Scarlet-chested Parrot.
2. Opaline Blue
   Scarlet-chested Parrot.
3. Pair of Scarlet-chested Parrots.
4. Pair of White-fronted Blue
   Scarlet-chested Parrots.

1. Sea Green Scarlet-chested Parrot.
2. Pied Scarlet-chested Parrot.
3. Cinnamon Blue Scarlet-chested Parrot.
4. White Scarlet-chested Parrot.
5. Skyblue split Cinnamon Scarlet-chested Parrot.

Page 29

1. Young cock Red-bellied or Full Red-fronted Scarlet-chested Parrot.
2. Par Blue Scarlet-chested cock.
3. White-fronted Blue Scarlet-chested cock.

1. Cinnamon Blue Scarlet-chested Parrot.
2. Golden Yellow and Ivory Scarlet-chested Parrots.
3. Cinnamon Scarlet-chested Parrot.
4. Sea Green x Cinnamon Scarlet-chested Parrot.
5. Cock Pied Scarlet-chested Parrot.
6. Cinnamon Scarlet-chested Parrot.
7. Pair of Red-bellied Scarlet-chested Parrots (cock on right).

ELEGANT PARROT
Neophema elegans

This is the first in the group of four Neophema species that are predominantly green in colour. These four, namely the Elegant, Blue-winged, Rock and Orange-bellied Parrot are not as popular as the Neophemas mentioned previously. They are mainly found in the aviaries of specialist breeders (with the exception of the Orange-bellied Parrot).

As mentioned earlier, the Elegant can bomb other Neophemas and is therefore untrustworthy if housed with any other variety of Neophema. Cock Elegant Parrots are also aggressive among themselves. This 220mm (8.6 inches) long Neophema normally broods up to six young in the nest and incubation lasts 18 - 24 days.

Above: Elegant hen.

Right: Young Lutino Elegant in the nest.

Below: Cock Elegant with the red 'thumb' mark on lower abdomen.

Sexing
Most cocks carry a red vent (thumbprint). The cock has a more pronounced frontal band than the hen. The lores on the cock are also a deeper, brighter yellow than those of the hen. The royal blue on the cock's wings travels under the wing and is strong in colour. Under the wing of the hen the blue tends to fade and merge into green. The underwing stripe is usually absent in both mature sexes. However, if the young are inspected before they leave the nest, hens often show the white stripe under the wing. Surgical sexing will of course remove any doubt and is recommended.

Mutations

Cinnamon
This is a sex-linked mutation.

Lutino
This bird is being bred in Australian aviaries. In the Lutino all blue becomes white. It is recessive.

Pied
This recessive mutation is becoming more available.

1. Cinnamon Elegant cock and Normal hen.
2. Pied Elegant Parrot.
3. Young Lutino Elegant in the nest.

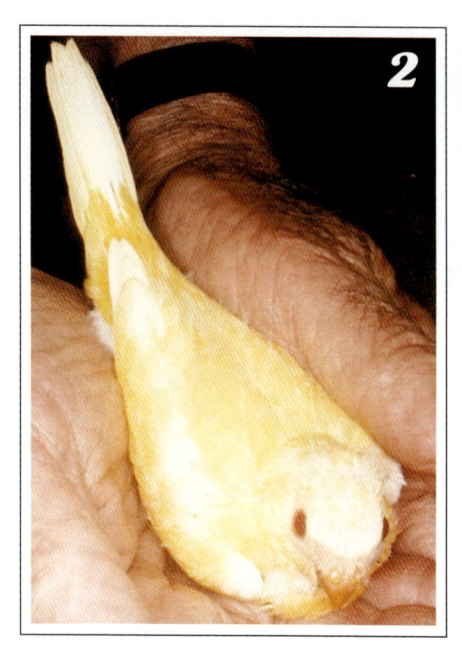

1. The Elegant Parrot shows more yellow on the chest than the Blue-winged Parrot.
2. Lutino Elegant Parrot.
3. Pair of Lutino Elegant Parrots.

Above: Blue-winged hen.
Right: Blue mutation.
Below: Blue-winged cock.

BLUE-WINGED PARROT
Neophema chrysostoma
Other names: Blue-banded Parrot.

The Blue-winged Parrot is quiet and peaceful and although breeding results often fluctuate from season to season, best results have been obtained through colony breeding rather than as single pairs.

With the exception of the Orange-bellied Parrot, of which little is known in aviculture, the Blue-winged Parrot is the poorest breeder of the genus. Breeding in the wild in Tasmania, south-eastern South Australia, Southern Victoria indicates that this bird dislikes hot climates. Overheating should therefore be carefully monitored with this species.

Sexing
The blue frontal band is less pronounced in the hen. The lores of the hen are not as bright or yellow as on the cock. The blue on the wing is duller in the hen than the cock and the blue on the underwing coverts on the hen is duller and quickly merges into olive green. Young hens tend to show the underwing stripe before they leave the nest.

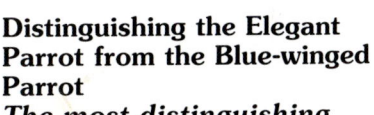

Distinguishing the Elegant Parrot from the Blue-winged Parrot
The most distinguishing points illustrated are:-
The Elegant has a double frontal band, the lower band being dark blue, and the upper band pale blue. Also, the band extends above and behind the eye.

The frontal band on the Blue-winged is a more uniform dark blue and does not extend behind the eye, as seen on the Elegant.

The Blue-winged is generally greener and has more blue on the wing than the Elegant.

The Elegant tends to show a brighter yellow on the chest.

Mutations

Blue Blue-winged
This mutation was originally bred in Australia, although it is still a rare mutation. It is recessive.

Page 35

ROCK PARROT
Neophema petrophila

The Rock Parrot is only found in aviaries in Australia. Many people find this bird difficult to breed.

Rock Parrots do not eat a lot of seed. They should be fed mainly vegetables, fruit, seeding grasses, Arrowroot™ biscuit etc. as described in general *Feeding*. In the wild they eat a lot of Pig-face, a wild succulent that grows on beaches. If they have no choice they will eat only seed and as a result, tend to become overweight. This will reduce their breeding potential and their life span.

Sexing
The dark blue frontal band is narrower on the hen than the cock, otherwise, they are similar. Surgical sexing is a must for this species.

Breeding
Rock Parrots have a different breeding habit to all other Neophema species. With other Neophemas you will find that if the young leave the nest slightly early and you put them back in the nestbox, within a short period of time they will be out again. You just cannot get them to stay in the box. The young Rock Parrots, however, continuously go back to the nestbox and often roost in the boxes at night, even when the hen is incubating a second clutch. They are very much a community bird and should be kept as such. They are active and enjoy playing with and chasing each other. Best breeding results are obtained when these birds are housed in colonies. Three pair colonies have proved extremely successful. By housing them in colonies this will tend to reduce obesity due to constant activity and exercise.

Rock Parrots will use normal nestboxes. It is not necessary to provide them with the rock crevices they use in the wild.

1. Rock Parrots are hard to sex visually.
2. Hen Rock Parrot.
3. Rock Parrots will breed in nestboxes. Rock crevices are not required in aviaries.
4. Cinnamon Rock Parrot.

Mutations

Yellow
Yellow Rock Parrots have been bred, however, no further details are known.

Cinnamon
This recessive mutation has red eyes.

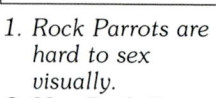

ORANGE-BELLIED PARROT
Neophema chrysogaster

The Orange-bellied Parrot in appearance is a typical member of the genus *Neophema* with many characteristics common to most others of the group. It is a small, slender, basically green parrot approximately 200mm (8 inches) in length and weighing about 55 grams.

Almost unknown in captivity, past efforts to maintain and breed the Orange-bellied Parrot have met with very little success and prior to 1986 the only successful captive breeding in Australia was recorded by the late Fred Lewitztea of Adelaide, who in 1973 reared two and in 1974 reared one young to independence. Numerous other attempts have always ended in failure.

With fewer than 150 birds remaining in the wild, it was essential that immediate action should be taken if the species was to be saved from extinction. Management initiatives were undertaken to protect and improve the habitat available to the birds both in their breeding range in south west Tasmania and on the mainland, especially in Port Phillip Bay in Victoria where important overwintering habitat occurs.

The Orange-bellied Recovery Team of Scientists from the Commonwealth and states of Victoria, Tasmania and South Australia with representatives of the Royal Australian Ornithologists Union (now Birds Australia) and the International Council for Bird Preservation, under the Chairmanship of J Forshaw, recommended in 1985 that a captive breeding programme be established and be carried out by the Tasmanian National Parks and Wildlife Service.

The programme commenced in 1986. In the first year (1986/1987), both of the two hens held went to nest and four young (3 and 1) were reared to independence. In the second year, three of four hens held, nested and eight young (4, 3 and 1) were reared to independence. The pair which reared one young let a further two young die in the nest. In 1988/89, all seven hens being held, nested and 22 young were reared to independence.

In summary, in captivity, most Neophemas are ready breeders. Through our experience we have found that the Orange-bellied Parrot is no more difficult than any others of the genus.

It is most important that both the captive breeding programme and the wild management programme should succeed in order to save this magnificent little parrot for the future.

Cock Orange-bellied Parrot.

BOURKE'S PARROT
Neopsephotus (Neophema) bourkii

As a result of findings from the research program, 'Biochemical Systematics of Parrots' carried out jointly by the Museum of Victoria, the CSIRO and the Australian National University, the Bourke's Parrot has been found to have no close relationship with the Neophema group.

Consequently the bird has been re-classified to its own genus - *Neopsephotus (Neophema) bourkii*.

The Bourke's Parrot does not have green as its basic plumage and colour, it is also not known to have hybridised with any Neophema species. This trait makes it the most suitable bird for sharing an aviary with Neophema Grass Parrots. This bird, with its subtle colourings of pink, blue and brown is well established in aviaries worldwide. Its willingness to breed also makes it very popular with beginners.

Bourke's Parrots are not very active during the day. However, at dawn and dusk they can be found fluttering and twittering around the aviary.

1. Cinnamon Bourke's Parrot hen.
2. The hen Bourke's Parrot lacks the blue frontal band.
3. New coloured Bourke's Parrot not named as yet.
4. Cream Bourke's Parrot hen.
5. Rose Bourke's Parrot.
6. (left to right). Cream, Normal, Pink and Cream Bourke's Parrot chicks.
7. Pink, White and Rose Bourke's Parrots.

1. Cream Bourke's Parrot hen.
2. Golden-mantled Bourke's Parrot.

1. Rose Bourke's Parrots.
2. Black-headed Golden-mantled Bourke's Parrot.
3. Yellow Bourke's Parrot (not a true Lutino mutation).
4. Pink Bourke's Parrot (known as Yellow Rose in Europe).
5. Bourke's Parrot - not named as yet
6. Pink mutation hen (left) Cream cock (right).

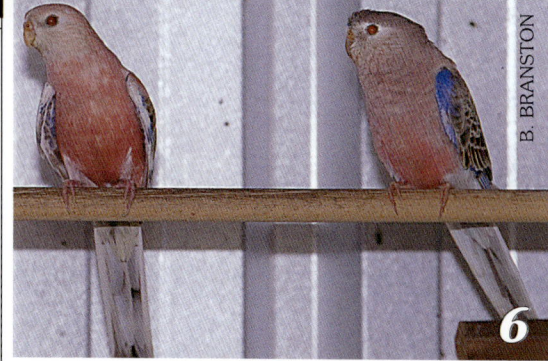

A handsome and subtly coloured Bourke's Parrot cock.

Their larger eyes indicate their preference and suitability for moving around in poor light. This after-dusk activity is the only reason some finch breeders do not house Bourke's Parrots in their aviaries, as they can have a disturbing effect on roosting finches.

Generally, these birds are very peaceful, although an eye must be kept open for the odd rogue bird.

Their general management, feeding and breeding has been well covered in the preceding chapter. Generally, up to seven young can be expected in the nest.

Sexing

The cock shows a blue brow, the hen a creamy white brow. As soon as one blue brow feather is seen on a young Bourke's Parrot you can be sure it is a cock.

Mutations

Cinnamon (Isabel)

This bird is referred to as an Isabel Bourke's Parrot overseas which may be a better name as it is recessive in its mode of inheritance. Splits are therefore available in both cocks and hens. The cock tends to be darker than the hen and shows the familiar blue brow. Adults retain red eyes. A true sex-linked Cinnamon may also occur but has not been properly identified.

Cream

This lovely bird is referred to by some as yellow. The amount of cream markings can vary on some birds. Like the Cinnamon, the Cream has red eyes and is recessive. The cock shows the blue brow, absent on the hen.

Rose

This must be one of the most beautiful Neophema mutations available. The photographs shown should speak louder than words but it is unfortunate that when photographed, the rich salmon pink colour often looks paler on these birds. This mutation is sex-linked. There is difficulty in sexing Rose Bourke's Parrots. There is no blue brow on the cock, however they are often a darker colour on the face than on the hen. They also develop increasing grey tones on the face, head and shoulders as they get older.

Pink

These striking birds are bred by crossing Rose with Cream Bourke's. The Pink Bourke's Parrot has red eyes. Pink Bourke's are sexed in the same way as the Rose Bourke's Parrot.

Red-eyed birds often dislike intense bright conditions. Red-eyed Bourke's Parrot mutations are no exception and shaded or darker aviaries have produced the best results.

Pied

Bourke's Parrots showing pied type markings in one way or another have been produced. Unfortunately, results in the second generations have been disappointing and this mutation still has to be 'fixed'.

Black-headed Golden-mantle - Genetic inheritance not known.
White Golden-mantle - Genetic inheritance not known.
Lutino - This is a sex-linked mutation.
White - Unsure of genetic inheritance.

The Psephotus Grass Parrots

Pair of Mulga Parrots

K. WILSON

Introduction

The Psephotus Grass Parrots can be divided into three groups:-
- **Red-rumped** *Psephotus haematonotus* and **Mulga** *Psephotus varius* Parrots.
- **Hooded** *Psephotus chrysopterygius dissimilis* and **Golden-shouldered** *Psephotus chrysopterygius* Parrots.
- **Red-vented Blue-bonnet** *Northiella (Psephotus) haematogaster haematorrhous*, **Yellow-vented Blue-bonnet** *Northiella (Psephotus) haematogaster haematogaster* and **Naretha Blue-bonnet** *Northiella (Psephotus) haematogaster narethae* Parrots.

As in the *Neophema* genus many of this group are becoming increasingly popular. Some, such as the Red-rumped and Hooded Parrots, also have their mutations. At the time of writing this revision, we believe, in Australian aviaries, there are estimated to be 22 mutations of Red-rumped Parrots, which are continuing to grow. Red-rumped Parrots are becoming more popular as show birds due to their many mutations and colour variations. Although not large birds, all members of this group are aggressive and especially spiteful to their own kind. One pair per aviary is a definite rule for Psephotus Grass Parrots.

Hooded and Golden-shouldered Parrots must rank among the most beautiful of parrots available to aviculturists today. The Blue-bonnet Parrot is more colourful than many realise. The pretty Mulga Parrot, with splashes of yellow and red on green brings us to the Red-rumped Parrot, maybe not as brightly attired as the rest, but certainly full of the one common factor that binds this group together, and that is character. When you own Psephotus Grass Parrots you will soon learn that they are all individuals - nosy, cheeky, aggressive, in a nice sort of way and arrogant. For this trait alone they make their presence in a collection worthwhile.

Management

Points made under *Management* and *Feeding* of the Neophema group also apply to the Psephotus group, the exception being the swapping around of cocks and hens in the breeding season. This practice would possibly have fatal results with any of the Psephotus family.

Housing

Let it be repeated, house one pair per aviary. No other birds should share their accommodation. Ideally, solid partitions should divide the flights, as this will avoid bickering and fighting between neighbours. Double wire essentially is the minimum division that should be considered for all Psephotus especially Blue-bonnet Parrots. Being able to hear but not see others of the genus most certainly stimulates the cock to come into breeding condition, ready to defend his mate and his nest site. Aviaries between 750mm (2.5 feet) to 1 metre (3.3 feet) wide x 2 - 3 metres (6.6 - 10 feet) long and at least 2 metres (6.6 feet) high will adequately cater for the needs of these birds. A solid shelter at least 1 metre (3.3 feet) deep similar to shelters mentioned under the Neophema group *Housing* should provide shade and a weatherproof retreat. Many breeders now completely cover the roof of the flight with a suitable transparent material. This eliminates all weather and disease problems that can be passed through a wire roof from wild birds and helps reduce cat

*Above left:
Australian breeding complex, housing Psephotus Grass Parrots.
Above right:
Rear ventilation doors.*

and hawk problems. Ample sunshine will penetrate the aviary through the front wire, providing of course the aviaries are facing north to north-east. The advantages of concrete floors are well known - easy to clean, helps control worm infestations, keeps out mice etc. If dirt floors are used, make sure tin or concrete walls are sunk at least 450mm (1.5 feet) into the ground all around the aviary to prevent vermin from digging in. Cover the roof completely with clear fibreglass or similar - dirt floors are one thing, muddy floors are quite another. To reduce problems with fallen seed, food scraps and worm eggs, dirt floors must be kept as dry as possible. Paving stones are also used successfully on aviary floors.

It should go without saying that a safety corridor or individual safety doors must be used. It is really inexcusable, if you think about it, to let a bird escape from your aviary.

Each aviary should contain two main natural perches (Eucalyptus etc.), one in the shelter and one near the front of the flight. An aviary full of perches can be hazardous to both the birds and owner and also encourages the birds to hop from one end of the aviary to the other rather than fly. Birds get maximum exercise by flying the full flight. It is therefore a good idea to have food and water receptacles below the perch level (not directly under of course). This means the bird must fly down to feed and drink and back up again to the perch.

Feeding

The *Feeding* section under the Neophema group is applicable for the Psephotus group. Please note the importance of calcium requirements. There is not much more to add except that under the specific species section that follows, various other foodstuffs such as Madeira cake and peas will be suggested. Your birds should be offered all such items to see what they will and will not eat. Vitamins and minerals in the drinking water are positively no substitute for a good varied diet.

Breeding

Bobbing up and down, strutting and hovering around the hens, feeding each other etc. are all signs of courtship displays. Individual breeding requirements are discussed under the relevant species.

Overall, it will be found that Red-rumped, Mulga and Blue-bonnet Parrots have basically the same management and breeding requirements. Hooded and Golden-shouldered Parrots are both similar in their requirements, also.

RED-RUMPED PARROT
Psephotus haematonotus
Other Names: Grass Parrot, Grassie

Pair of Red-rumped Parrots.

These charming character birds must contend with the Eastern Rosella and Cockatiel as the ideal beginner's parrot. This parrot is becoming more popular due to its large variety of colour mutations and attractiveness of the colourations within these mutations. The Red-rumped Parrot is extremely hardy and can be kept under varying weather conditions. It is a pity that out of the above mentioned three species, the Red-rumped and Eastern Rosella are aggressive birds. Many beginners have only one aviary and are tempted to house different species together. Invariably problems arise between these two species and other birds. It must therefore be stressed again that they should be housed in single pairs. Hence, when you are starting out in keeping parrots, two aviaries 3 metres (10 feet) long x 1 metre (3.3 feet) wide is a far better proposition than one aviary 3 metres (10 feet) long x 2 metres (6.6 feet) wide. Red-rumped Parrots will do well in aviaries 750mm-1 metre (2.5-3.3 feet) wide x 2-3 metres (6.6-10 feet) long x 2 metres (6.6 feet) plus high. The general housing rules mentioned previously apply.

Feeding

Red-rumped Parrots should present no feeding problems to their owners. They usually eat everything given to them. Feed a seed diet similar to Neophemas, being French white millet, canary seed, grey-striped sunflower seed and hulled oats and occasional green feed during the non-breeding season. Very little sunflower seed is offered. During the breeding season the birds are fed sunflower seed, sprouted mung beans, greens, seeding grasses and an Arrowroot™ biscuit every day, as well as, the staple mix of canary and French white millet seeds. The birds have no problems due to being overweight and the increase in foodstuffs encourages them to begin breeding. The same technique is used with great success with Mulga Parrots.

1. Cock Cinnamon Red-rumped Parrot.
2. Lutino Red-rumped Parrots.
3. Hen Cinnamon Red-rumped Parrot.
4. Hen Platinum Red-rumped Parrot.
5. Cinnamon Opaline Red-rumped Parrot.

Page 47

1. UK Pied Red-rumped Parrots on the wire.
2. Australian Dominant Pied Red-rumped Parrot
3. Hen Blue Red-rumped Parrot.

1. Cinnamon Blue Opaline Red-rumped cock and hen.
2. Blue Red-rumped cock.
3. UK Cinnamon cock.
4. Cinnamon Blue Opaline Red-rumped cock.

Page 49

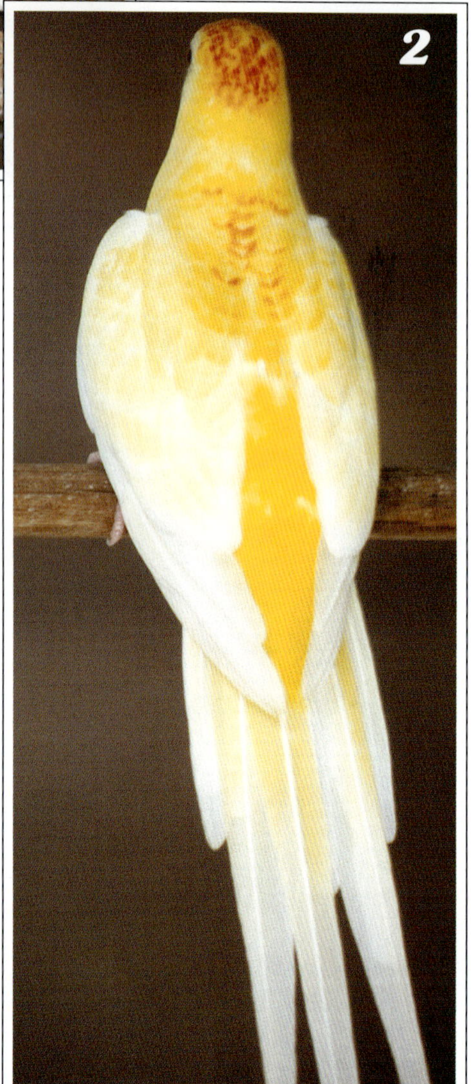

1. Normal Red-rumped cock
   and Platinum
   Red-rumped cock.
2. Lutino Opaline
   Red-rumped hen.
3. Albino Red-rumped Parrot.

Page 50

1. Cinnamon Blue
   Red-rumped hen and cock.
2. Opaline cock.
3. Cinnamon Opaline hen.
4. Lutino Opaline cock.

Breeding

The recognised breeding season in Australia is from June to December, but like many of our aviary bred parrots, if it suits them they will breed. It would therefore be safe to say that they could breed any time between May and January.

Nestlogs or nestboxes can be used and should be about 150mm (6 inches) square by 450mm (1.5 feet) high. Wire or timber ladders can be attached inside the nestboxes. Logs are preferably hung just off the vertical to reduce the chances of the parents jumping straight down onto the eggs.

1. Blue Opaline Red-rumped Parrot.
2. A beautiful cock Blue Red-rumped Parrot.
3. Cinnamon Opaline and Opaline Red-rumped Parrots.

1. Opaline Red-rumped cock.
2. Lutino Platinum
   Red-rumped cock.
3. UK Pied cock.

1. Platinum Red-rumped hen.
2. UK Cinnamon hen.
3. Cinnamon Opaline Red-rumped Parrot.
4. Blue Pied Red-rumped hen and UK Pied Red-rumped Parrots.

*Above left:
UK Cinnamon cock.
Above right:
Lutino Opaline cock.*

A compatible pair of Red-rumped Parrots is the key to success. These are always sitting and feeding together, preening one another etc. It is not uncommon for them to raise four broods in a season, two broods being normal. The cock feeds the hen while she broods alone. Four to seven (usually five) eggs are laid per clutch and incubation lasts approximately 20-25 days. The young usually hatch on the same day. Certainly all are hatched over a 48 hour period. A further four to five weeks and the young will be leaving the nest. They are usually independent two weeks after fledging. Young birds can be left longer with the parents but the parents must be watched, for any signs of aggression. A careful eye must be kept on the hen. She often resents the young (especially young hens) when she begins her second brood. The cock usually presents no problems.

The young can be sexed in the nest. The young cocks are brighter in colour and red feathers can be seen coming through on their rumps.

Red-rumped Parrots will breed from between 12 and 18 months of age. Cocks tend to mature faster than hens. All in all the Red-rumped Parrot is a no-problem bird and very willing to breed.

Mutations

There are some 22 mutations of Red-rumped Parrots known to be available in Australia. Also, Australia is the most advanced country in the world in the breeding of Red-rumped mutations.

There are many variations of the Blue Red-rumped, Opaline Red-rumped and Pied Red-rumped Parrot including Dominant Pieds. The accompanying photographs cover just some of the ever-increasing range of colour mutations in Red-rumped Parrots.

Cinnamon

Genetically, there are two Cinnamon varieties in the Red-rumped Parrot. One is sex-linked, the other is recessive and should be called Isabel. They cannot easily be distinguished visually. Recessive Cinnamons (Isabels) sometimes show a creamier coloured beak and lighter coloured legs than the sex-linked variety. Cinnamon Red-rumped Parrots do vary in colour tones. Some are a dull biscuit colour, others lean towards a bright lemon shade. The more prevalent Cinnamon appears to be the sex-linked variety. The cock Cinnamon shows a red rump, which is absent on the hen.

Also bred in Australia are imported UK Cinnamon Red-rumped Parrots which are now developing further variations of this mutation.

Lutino

This is another sex-linked mutation. The cock is bright yellow with a few paler to white feathers on the wings and tail. He still retains the red rump which looks very striking on yellow. The hen is paler than the cock. Like all Ino's they have red eyes. The Lutino is now being bred with Opaline, Platinum and Blue Red-rumped Parrot mutations.

Golden Fallow Red-rumped Parrot.

Blue

The Blue Red-rumped Parrot is now being bred with Cinnamon, Lutino, Pied, Platinum and Opaline mutations. The Blue mutation is recessive, so both cocks and hens can therefore be split for blue.

The cock has a blue head and darker blue shoulders, a white rump and blue tail. The rump in the hen is blue, with no white at all.

Blue combined with Lutino will produce Albino, with Black-eyed Yellow will produce White and with Pied will produce Blue Pied. Cinnamon Blue should not be called Silver as they are blue and fawn in colour without any silver. The real Silver will one day be bred.

Pied

In Australia now, we have two different pied mutations. The first is the typical pied, called Australian or Dominant Pied. The second is the imported UK or Olive Pied. This mutation is recessive and whilst it produces pied markings, it also alters the shade of the bird's body colour from bright green to olive green, hence the name Olive Pied. Its effect on the Blue mutation has not been thoroughly studied as yet.

Opaline

A sex-linked mutation, this attractive bird will show the many variations possible with this mutation. They have already been bred with Blue, Cinnamon and Lutino mutations. There is enormous scope for the development of further mutations in the Red-rumped Parrot over the coming years.

MULGA PARROT
Psephotus varius
Other names: Many-coloured Parrot, Varied Parrot

The Mulga Parrot feeding and housing requirements are generally the same as for the Red-rumped Parrot. Less common in aviaries than the Red-rumped, the Mulga Parrot is not as aggressive, however, one pair per aviary still produces the best results.

Breeding
Mulga Parrots can be double brooded but they are not as prolific as the Red-rumped. A normal clutch is four to six eggs, however, the Mulga Parrot has been known to produce clutches of up to 10 chicks. Only the hen broods. Incubation lasts 20-25 days and all young hatch within a 48 hour period. On fledging, the chicks are much more mature looking than Red-rumped chicks. They have more colour, their tails are longer and they are less flighty. They stay in the nest

Above: Pied Mulga Parrot cock.
Right: Mulga Parrot hen.
Below: Cinnamon Yellow Mulga Parrot.

about a week longer than the Red-rumped. When the young are being fed it will be noticed that the larger food particles such as sunflower seed, sprouted mung beans, Arrowroot™ biscuit etc. disappear very quickly. This is because it is easier, quicker

Mulga Parrot cock.

K. WILSON

and therefore less tiring for the parents to fill their babies' crops with these foodstuffs rather than collect small millet seeds and the like.

Sexing

Young Mulgas can be sexed in the nest. Cocks show yellow feathers and hens red feathers on the shoulder.

Mutations

Pied and Cinnamon Yellow mutations are bred in Australia. There has been mention of other mutations which the author has not seen personally. Due to the lack of recorded information on these mutations, further details are unavailable at the time of writing.

Pair of Mulga Parrots.

Page 59

BLUE-BONNET PARROT
Northiella (Psephotus) haematogaster

Since the first edition of this book was published, the Blue-bonnet Parrot has been reclassified into its own genus. (The Taxonomy and Species of Birds of Australia and its Territories, 1994, L. Christidis and W. Boles.)

It is generally accepted in aviculture that there are three subspecies of the Blue-bonnet Parrot:-

- **Yellow-vented Blue-bonnet**
  Northiella h. haematogaster
   - Red abdomen with yellow vent (Nominate subspecies).
- **Red-vented Blue-bonnet**
  Northiella h. haematorrhous
   - Red abdomen with red vent.
- **Naretha Blue-bonnet**
  Northiella h. narethae
   - Yellow abdomen with red vent.

The Naretha Blue-bonnet is much smaller than the Red-vented and Yellow-vented Blue-bonnets and becoming more readily available to aviculturists.

Blue-bonnet Parrots are one of the most misunderstood birds available to aviculture today. They are regarded as noisy, drab, extremely aggressive and poor breeders. From the previous sentence, only the aggressive tag really suits. Most of our information is passed down over quite a considerable time and little or no updating due to more advanced aviary management is available.

Blue-bonnets certainly are aggressive birds but probably no more so than the other members of the Psephotus group. Although Blue-bonnets have some raucous calls, I doubt if they could be called noisy, when compared to Lorikeets for example. Although from a distance, Blue-bonnets appear mainly brown, on closer inspection they are very pretty with splashes of blue, red and yellow.

B. BRANSTON

Yellow-vented Blue-bonnet Parrot.

Red-vented Blue-bonnet Parrot

B. BRANSTON

Housing

Blue-bonnet Parrots, if housed sensibly and given a suitable nesting receptacle should, and usually do, provide a rewarding experience. Blue-bonnets must be housed as single pairs and mixed with no other species of birds. Double wire or solid wall partitions are essential. Aviaries should be completely roofed. An aviculturist's greatest enemy is the weather. Birds do not need to be rained on and if your aviaries face in a northerly direction (as they should) ample sunlight will be provided. If you want to wet your birds, install a sprinkler system. At least one third of the aviary roof should be steel, aluminium or fibro to provide shade and shelter. The remainder of the flight should be covered with clear fibreglass sheeting or similar.

Feeding

A simple seed mixture of mixed millets and canary seed with a little sunflower seed is all that is necessary as a basic diet. Green seeding grasses should be supplied whenever available as well as fresh corn on the cob, apple, endives and soaked and sprouted seeds. During the breeding season all of the above foods should be increased and Arrowroot™ biscuit supplied, especially while chicks are in the nest.

Breeding

Blue-bonnets nest early in the season, usually late July in Australia. The birds are fairly secretive about their nests and require a log of fairly strict dimensions before they will accept it. A true pair should show interest in a nestlog of about 600mm (2 feet) high with an internal diameter of 150mm (6 inches). The entrance hole should either be a small natural knot hole or a hole cut about 100mm (4 inches) down from the top of the log and measuring about 50mm (2 inches) in diameter. Logs should be hung almost upright and filled to a depth of about 75mm (3 inches) with decayed wood pulp, crushed termite nest or similar. The hen usually sits very tightly and disappears into the log on sighting anyone. Nestboxes are accepted by some pairs but a suitable log usually has more appeal. Nestboxes should be about 200mm (8 inches) square by 600mm (2 feet) high.

The usual clutch is about five eggs but some pairs will lay up to seven. Only the hen broods and incubation lasts 20-25 days. The young remain in the nest for about five weeks. About two weeks after fledging the young should be independent. A watch must be kept on parents for any signs of aggression towards the young.

For obvious reasons, Red-vented and Yellow-vented Blue-bonnets should not be interbred.

Mutations

White, Cinnamon and Pied mutations of Blue-bonnets exist in Australian aviaries, however, further details on their genetic inheritance are not available at the time of writing.

Above:
Red-vented
Blue-bonnet Parrot.
Right:
High Red-fronted
Blue-bonnet Parrots.
Below:
Naretha Blue-bonnet
cock.

B. BRANSTON

HOODED PARROT
Psephotus dissimilis

The Hooded Parrot, over recent years, has become a highly prized species within Australian aviculture. However, obtaining a compatible pair is not always easy.

If compatible pairs cannot be purchased, then unrelated, uncoloured birds should be selected. Hopefully these will have paired up when fully coloured at around 12-16 months of age. It is often difficult to find compatible partners for single mature Hooded Parrots.

Both the Hooded and the Golden-shouldered Parrot are growing in popularity every day. The number of people who have problems with these birds is quite alarming. For these birds, the old saying that 'there are no hard and fast rules in aviculture' must be completely ignored. Unless the strict basic rules are adhered to, success with these birds will be very limited.

Both Hooded and Golden-shouldered Parrots should be wormed at least twice a year.

1. Hooded Parrot cock.
2. Cock Pied Hooded Parrot.
3. Cinnamon Hooded Parrot cock.
4. Hooded Parrot mutation with red wings.
5. Dark coloured Olive Hooded Parrot - five months old.

Birds should not be wormed during the moult or all growing flight and tail feathers will often fall out. However, there appears to be no permanent damage done, as the next feather growth is usually normal.

Housing

There should be only one pair of Hooded Parrots housed per aviary. No other birds should be kept in the same enclosure. Aviaries must be fully covered, so the birds, especially the young, are not subjected to inclement weather. An ideal sized aviary is 3 metres (10 feet) long x 1 metre (3.3 feet) wide x 2 metres (6.6 feet) high. Any longer is a waste of time, material and effort. Longer flights allow such swift flying birds time to build up momentum which can prove fatal to young birds or night flyers on sudden contact with the end wire.

A solid shelter must make up at least one third of the total aviary length to provide shade and protection. The remainder of the flight should be completely roofed with a semi-transparent or transparent material.

Solid or at least double-wired partitions should divide each aviary containing these or other species. Concrete floors should be covered to a reasonable depth with river sand or the like. These birds love to dig. Other points mentioned in this book, such as direction of aviaries and points on cleanliness, draught and damp etc, of course apply.

Feeding

Many food items already mentioned will be adequate for Hooded and Golden-shouldered Parrots. Seed mixtures should be kept simple eg. canary seed, French white millet and hulled oats. Grey-striped sunflower seeds and hulled oats are given at about half the amount of the small seed mix. Greens are greatly appreciated especially green peas. Seeding grasses should be given when available. A mixture of sprouted seed, described under general *Feeding* of Neophemas is ideal.

Calcium is necessary for the well-being of adults and young and these birds seem to need a copious supply. Cuttlefish bone and grit should always be available. If the cuttlefish is not being touched, replace, reposition or break it up. Alternately, use calcium blocks in the drinking water, as described under general *Feeding* of Neophemas.

1. Pair of Hooded Parrots.
2. Cinnamon Hooded Parrot hen.
3. Pied Hooded Parrot.

Below: Hen Hooded Parrot at entrance of her nest.
Right: Green-Yellow Pied Hooded Parrots.

Breeding

In their natural habitat the Hooded Parrot breeds in the dry season. In captivity these birds will usually breed between February and September, in Australia. Courtship displays in both Hooded and Golden-shouldered Parrots are similar. The cocks raise their frontal head feathers to form small crests. They puff their chests out and strut and flutter around their hens.

Hooded and Golden-shouldered Parrots lay an average of four to six eggs per clutch. Incubation lasts approximately 20-25 days. Only the hen incubates and is fed by the cock. The young leave the nest at about five weeks of age and are independent anywhere between 10 and 20 days afterwards. The young should be removed as early as possible - as soon as they are seen eating from the seed dishes. Aggressive parents, especially the cocks, must also be carefully monitored. Young cock Hooded Parrots can be identified by their bright blue cheek patches. They are fully coloured by 18 months of age, however, cocks have successfully bred while still colouring up.

Although they nest in termite mounds in the wild, in the aviary, nestboxes are sufficient. The nest chamber should be just big enough for the hen and her chicks and be insulated. What has proved a very successful nestbox design has an outside box measurement of 300mm (12 inches) long x 200mm (8 inches) wide x 200mm (8 inches) high. The internal nest chamber measures 150mm (6 inches) long x 150mm (6 inches) wide x 150mm (6 inches) high without any ventilation holes, as the entrance tunnel or spout provides adequate air. The gap between the two boxes is insulated with polystyrene or similar insulating material. Each box has its own separate lid and a spout 150mm (6 inches) long passes through the first box into the inner chamber. PVC piping, grooved inside to provide grip for the birds, is adequate for the spout. The spout should be just big enough for the birds to squeeze through, about 60mm (2.4 inches) in diameter. Although this box is very successful for Hooded Parrots, I do recommend wherever possible to use heated nestboxes, especially south of Sydney and the southern Tablelands, New South Wales and cooler climates.

These birds come from areas in which the temperature is seldom below 30°C (86°F) at any time of the year, and their tunnels in

Cross Section Of Nesting Box

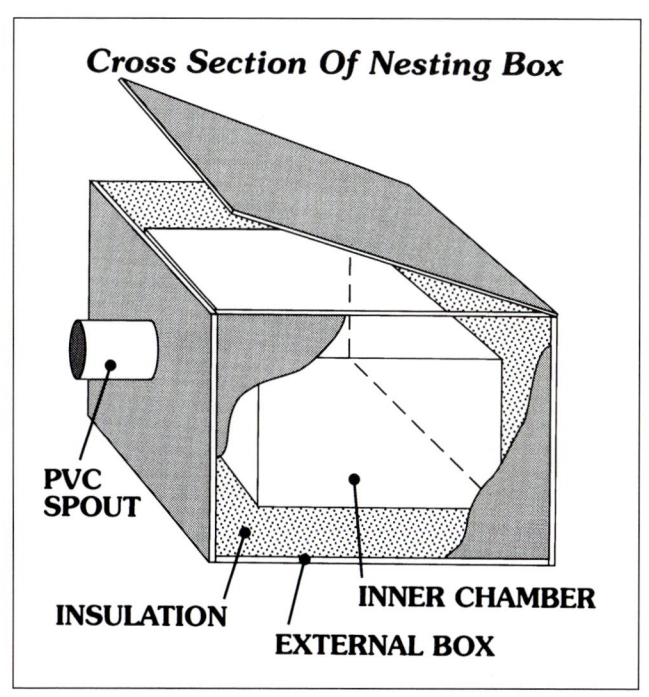

- PVC SPOUT
- INSULATION
- INNER CHAMBER
- EXTERNAL BOX

termite mounds are the only method of air circulation to the nesting chamber. It is for this reason that wild Hooded Parrots can leave their nests for long periods of time. The heat generated within live termite mounds is sufficient to keep the eggs warm. Finely crushed termite mound (tree type) or decayed wood pulp is used as nesting material. This is soaked in water for a few days before being squeezed and placed damp in each nestbox to a depth of about 50mm (2 inches). Do not be alarmed if most of the nesting material is removed. The nestboxes should be cleaned and the material replaced after each clutch. The nestboxes can be positioned at any height from 1 metre (3.3 feet) off the ground. The birds seem more at ease if the boxes are facing the front of the aviary as they can watch for any approaching danger.

In conclusion, check your birds morning and night and never be afraid to check nestboxes. Enter your aviaries regularly and earn the trust of your birds. If you want them to be good to you be good to them. If a bird becomes ill, treat it immediately or take it to an avian veterinarian - don't wait to see if it will improve. Keep detailed records of all happenings in your aviaries as this information will prove invaluable in later years.

Little more can be said except that these birds are generally not easy to manage (again there will be exceptions) and unless you give them what they need, not what you want, you will never be a successful breeder of Hooded and Golden-shouldered Parrots.

Mutations

Pied Hooded - Variations of Pied Hooded Parrots include Yellow, Green suffused Yellow and Green-Yellow.
Olive - No information on this mutation is available.
Red-shouldered - Possibly a colour variant.
Cinnamon - First bred in eastern Victoria, Australia, 1989.

Pair of Golden-shouldered Parrots

P. ODEKERKEN

Young cock Golden-shouldered Parrot.

GOLDEN-SHOULDERED PARROT
Psephotus chrysopterygius

Housing, **Feeding** and **Breeding** are as described for the Hooded Parrot.

Breeding (additional notes)

In cooler to cold climates, it is recommended that heated nestboxes are used for optimum breeding results in Golden-shouldered Parrots. Like the Hooded Parrot, in the wild, Golden-shouldered Parrots nest in live termite mounds which retain a constant temperature when the birds are absent.

The nestbox is completely filled with rotted wood material or very fine pine chips up to the entrance of the spout. Some breeders even make a light mixture of clay, plugging the spout up so that when the hen digs into the clay, as she does in a termite nest, it stimulates her into breeding condition. She will then possibly remove 80% of all the nesting materials prior to laying her clutch of eggs.

Mutation

The Fallow Golden-shouldered is the only known mutation.

Fallow Golden-shouldered Parrot.

Page 67

PARADISE PARROT
Psephotus pulcherrimus

The last confirmed sighting of the Paradise Parrot was by Mr C. H. Jerrard in the upper Burnett River area, southern Queensland, on 14 November, 1927.

Although unconfirmed sightings continue to filter through to Queensland authorities it is extremely doubtful that this beautiful parrot actually still exists in the wild, let alone captivity, today.

Below and left: Paradise Parrot.

Diseases of Neophema and Psephotus Grass Parrots

Cock Elegant Parrot

Introduction

Many of the disease problems encountered in these birds (eg. worms, mites and lice, bacterial and some viral diseases) can be prevented by sound management. The Poultry industry learned many years ago what aviculturists are finding today: appropriate quarantine procedures, aviary design, pest control, good diet and cleanliness are essential to maintaining a healthy collection. Sound management prevents disease entering the aviary and controls disease within the aviary.

There are several topics which will be discussed under the following headings:

- The Health Check of Newly Acquired Birds
- Quarantine of New Birds
- Stress and Disease
- Control and Prevention of Disease in the Aviary
- Worms and a Worming Programme including Average Weights of the Birds

The Health Check of Newly Acquired Birds

Newly acquired birds represent the most common disease danger to the aviary because some diseases are difficult to detect. It is necessary to properly examine any new bird before it enters your quarantine aviary. This examination starts at the head, checking the eyes, the nostrils, the feathers around the nostrils, the head, body, wings and especially the tail for signs of abnormal feathers, the vent for signs of pasting and the keel bone for 'going light'. The birds must be checked carefully for watery eyes and discoloured or malformed nostrils. These are signs of conjunctivitis and sinusitis. These diseases are especially prevalent in the Bourke's, Turquoisine, Elegant, Bluebonnet and Red-rumped Parrot. Left untreated for any period of time, these diseases often become impossible to cure. Both are symptoms of Chlamydiosis (Psittacosis). This is a particularly dangerous disease to birds as well as humans and you must be able to recognise its symptoms. Birds which have a prominent keel bone are said to be 'going light'. This has many causes, most of which are dangerous to your resident flock. Such a bird should not be purchased. It is very important to check the feathers. Many feather abnormalities seen in this group of birds indicate viral diseases such as Papovavirus or Psittacine Beak and Feather Disease. These diseases are usually not detected until the breeding season. It is advised not to purchase Hooded and Golden-shouldered Parrots during a moult.

The health check of the new bird is the first step in preventing disease from entering your aviaries. The quarantine procedure is necessary to detect any diseased birds which have passed the new bird health check. Some birds appear healthy but in fact may carry a disease. These birds are called 'carriers' and quarantine procedures allows us to detect such birds. Many of these diseases are stress related and within the six to eight week quarantine period most of these diseases will become apparent.

Quarantine of New Birds

Quarantine gives the birds time to acclimatise to their new surroundings, so that by the time they enter the aviary they are not in a stress situation. This reduces the chance of introducing disease to your aviary. During the quarantine period Neophema and Psephotus Grass Parrots should be treated routinely for worms and mites using Ivermectin or Panacur 25™. The only birds that can be housed together in the quarantine area at one time are those of a compatible species and from the same breeder. Any bird showing signs of illness must be immediately isolated. All healthy quarantined birds are allowed to enter the main aviary after six to eight weeks. The quarantine aviary is then disinfected and cleaned in preparation for the next new purchase.

Stress and Disease

Stress makes birds more susceptible to disease. All species in this group, except perhaps Red-rumped and Bourke's Parrots, are very susceptible to stress. Stress can be brought on from noise, vermin, such as mice and rats at night, an incorrectly designed aviary, unhygienic aviary management, incorrect diet during the breeding season or occurrences which affect the physical and psychological well-being of the bird. Stress control is a very important part of disease prevention.

Recognising Disease

Early recognition of disease is extremely important in their control. These smaller parrots succumb to the effects of disease more quickly than their larger relatives. Neophemas are particularly sensitive to the effects of disease due to their relative inability to conserve body heat. Observe and familiarise yourself with your birds' behaviour. Any changes which may signal illness will then be noticed earlier. Initial signs of an unwell bird are often subtle, as birds have the ability to hide their illness, this is known as the Preservation Reflex.

Observe whether your bird is less active than normal, fluffed up with both feet on the perch, rubbing its head on the perch, has a slight tail bob (this could also indicate hens are about to lay), has its head behind its wing, is sitting low on the perch, appears sleepy, not eating with the other birds, does not fly away when you enter the cage or is lethargic on the ground.

Or the bird may sneeze, display wet feathers around the head, a wet eye, a shut eye, a droopy wing, faeces stuck to the vent or tail feathers, or be unable to fly.

Once recognised as being ill, the bird should be moved to a hospital cage. This cage provides an opportunity for the bird to rest in a controlled environment, isolates the bird from the rest of the aviary and allows you to more closely observe the bird and begin any intensive treatment which may be necessary.

Control and Prevention of Disease

It is necessary for you to know what diseases are present in the aviary before a reliable disease prevention programme can be initiated. For this reason any ill or dead birds should be properly analysed. As well, a knowledge of those diseases most common to this group will help to prevent diseases entering the aviary.

At the time of purchase you can check for these diseases, and treatment can be given in quarantine for those diseases difficult to detect by examination alone. The health of expensive new birds can be thoroughly checked by your avian veterinarian.

Neophema and Psephotus Grass Parrots, being ground feeders,

are susceptible to diseases spread via droppings - notably worms (Roundworms, Hairworms and Gizzard worms), Chlamydiosis and bacterial enteritis. Psephotus Grass Parrots eat insects and even gastropods (snails/slugs) which can give Tapeworm or proventricular worms. Diseases such as French Moult (Psittacine Beak and Feather Disease, PBFD), Papovavirus and Megabacteria do not escape these groups of birds. These diseases are difficult to detect outside the breeding season. External parasites, such as Feather Mite, Feather Lice, Red Mite and Scaly-face Mite are a problem with the smaller Neophemas, particularly the Turquoisine and Scarlet-chested Parrots.

Prevention of Disease within the Aviary

The Neophema and Psephotus Grass Parrots year can be divided into the non-breeding, moulting and re-housing seasons. The disease prevention programme is designed around these seasons. The time of onset of the breeding and moulting seasons may vary from year to year. However, usually for this group, excluding the Golden-shouldered and Hooded Parrots, the breeding season in the southern hemisphere begins in late July/early August, and the moulting season occurs around January/February. The Golden-shouldered and Hooded Parrots may breed in August/September or in February depending on climatic conditions, and moult around late December and again after winter (July). Most bird keepers purchase new birds in late February (the re-housing season). It is best not to use antibiotics and other medications during the moulting and breeding season, and special attention should be given to re-housed birds.

Some diseases in this group occur more often in one season than another eg. worms occur after wet/humid weather. Chlamydiosis occurs more often during the re-housing season. PBFD becomes more visible after the moult. Your avian veterinarian can formulate prevention programmes taking this into account.

Worm Treatment

Worming is best done using the smallest size crop needle and a tuberculin (1 ml) syringe. The small calibrations of the tuberculin syringe gives a very accurate measurement of the small volumes of worm medicine needed for these small species. It is better to use smaller volumes so that reflux from the crop into the lungs is prevented. From 0.1 ml to 0.2 ml dosages are best for Neophema and Psephotus Grass Parrots. (See table for individual dose rates).

It is necessary to worm Neophema and Psephotus Grass Parrots at least three times a year. There will be deaths from obstructed and ruptured bowels if this group of birds is not routinely dewormed. The most appropriate times to deworm your birds are before the breeding season, after the breeding season and after the warm humid weather seen around December in Australia.

Worming Programme

Use a crop needle to dose against worms.
- Worm one month before the breeding season and repeat in three weeks.
- Worm the parents and chicks at time of fledging.
- Worm the parents and the young birds at the end of breeding season.

- In aviaries where a worm problem has been diagnosed it is necessary to repeat the worming process in early December and even repeatedly every three months throughout the year.

Refer to the list of the recommended dose rates for each species of Neophema and Psephotus Grass Parrot. (See disclaimer). It must be noted that some species can vary in their average body weight and the dose must be adjusted up or down accordingly. Ideally, the bird should be weighed before dosing to ensure the correct dose, but the following list can be used as a guideline when using safe medicines such as the wormers listed. There appears to be greater variations in the weight of Red-rumped Parrots and Red-vented Blue-bonnet Parrots than for the other species. The medication should be measured using the tuberculin syringe so that an accurate dose can be administered.

Ivermectins are now available in many different formulations and strengths. Each formula has a different dose rate so the reader should confer with their avian veterinarian.

Some formulas can be used effectively as spot-on treatments, all are unstable in water and are unreliable given this way. Ivermectins are excellent for the control of mites.

Oxfendazole has gained popularity as an in-water treatment for worms because it has no taste. Once again different formulas have differing concentrations. All are used in water at a concentration of 100-200 mg/L. If being administered by crop needle, repeat for three consecutive days.

Fenbendazole should be administered for three consecutive days to ensure effectiveness. This makes it either difficult or unreliable as it cannot be used in water.

Fenbendazole and oxfendazole should not be used during moulting as they may cause feather abnormalities.

Pyrantel is a common ingredient in dog and human wormers and is very effective against nematodes. See table for dose rates. Must be administered by crop needle.

Levamisole is a good old reliable. It has a bitter taste in water, however is very effective administered via crop needle - see dose rates.

Dose Rates for Deworming each Species

Species	Sex	Av Weight	Fenbendazole Dose 0.2ml/100gm Panacur 25™ 25mg/ml for 3 days	Levamisole Dose 0.1ml/100gm Nilverm™ Pig and Poultry Wormer 16mg/ml	Pyrantel Dose 50mg/ml Combantrim™ Dilute 1:10 with water 0.2ml/100gm
Blue-winged	M F	45-50g 45-50g	0.09-0.10ml	0.05ml	0.10ml
Elegant	M F	45-50g 45-50g	0.09-0.10ml	0.05ml	0.10ml
Turquoisine	M F	45-50g 45-50g	0.09-0.10ml	0.05ml	0.10ml
Scarlet-chested (Splendid)	M F	45-50g 45-50g	0.09-0.10ml	0.05ml	0.10ml
Bourke's	M F	45-50g 45-50g	0.09-0.10ml	0.05ml	0.10ml
Rock	M F	50-55g 55-60g	0.10-0.11ml 0.11-0.12ml	0.06ml	0.12ml
Hooded	M F	60-65g 55-60g	0.12-0.13ml 0.11-0.12ml	0.06ml	0.12ml
Golden-shouldered	M F	55-60g 50-55g	0.11-0.12ml 0.10-0.11ml	0.06ml	0.12ml
Red-rumped	M F	65-100g 55-85g	too variable	0.07ml	0.14ml
Mulga	M F	60-70g 55-70g	0.12-0.13ml 0.11-0.14ml	0.07ml	0.14ml
Blue-bonnet	M F	100-130g 100-140g	too variable	0.10ml	0.20ml

DISCLAIMER

Note Carefully: There have been little or no scientific studies documenting the safe use of these anthelmintics (worming medications) in Neophema and Psephotus Grass Parrots. The use of these drugs remains the full responsibility of the owner of the birds.

The recommended dose rates above have been found to be effective and safe in non-scientific field trials. It is important not to give any other drugs whilst using the above anthelmintics. If there are adverse effects contact your avian veterinarian for advice.

Publishers Note

In the revision of this book it was decided to delete a large portion of the Disease section due to the release of Dr Mike Cannon's ***A Guide to Basic Health & Disease in Birds*** which covers in greater detail the information contained in the deleted section. It is strongly recommended that this title be consulted in the diagnosis and/or treatment of sick birds. It must also be stated that if in any doubt regarding appropriate measures to take with a sick bird, seek the advice of a qualified avian veterinarian.

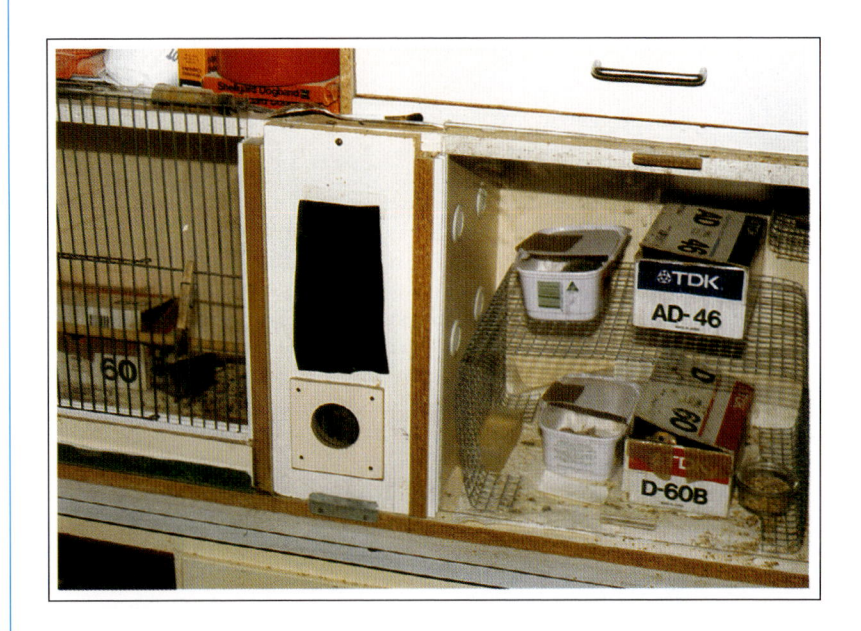

Right: The author's double brooder.
Below: The hot-air source of the brooder and hospital units.

Double Hospital Cage and Brooding Box
(As Used By The Author)

The total unit measures 1.04 metres (3.4 feet) long x 43cm (17 inches) high x 33cm (13 inches) deep and is made of laminexed chipboard for easy cleaning. Each cage is 43cm (17 inches) long leaving a width of 18cm (7 inches) in the middle to house the heat source. The roof of the unit is 25mm (1 inch) less in width than the base. This allows the perpendicular fitting of plastic tracks to allow operation of the front sliding covers.

One cage is used as a baby chick unit, the other houses older chicks to be weaned onto seed. The baby unit is thermostatically controlled. The warm air enters through four vent holes in the side. The temperature is maintained at 30-32°C (86-89.6°F). The chicks are kept in small boxes (as single birds or groups - from one nest). A middle wire shelf allows room for two tiers of boxes.

Page 75

The older chick unit does not contain a thermostat. Only one vent allows warm air to enter. The temperature in this unit is approximately 28°C (82.4°F). There is a tin tray at the bottom to facilitate cleaning. Perches, seed and water dishes are also provided. Both units contain a minimum-maximum thermometer, and have wire cage fronts with an inside sliding perspex front.

The motor housed between the units is an intake fan unit from an incubator. Heat is provided by two 60 watt globes. The 40mm (1.5 inches) hole at the bottom is covered in a fine gauze to filter the air. A 50mm (2 inches) x 100mm (4 inches) perspex hole allows checks to be made on the globes.

The hospital cage unit is identical. Unlike many hospital cages, this version does not force the bird or food and water dishes to be directly above the light bulb. The temperature is thermostatically controlled and is kept at about 35°C (95°F).

Handrearing

Sometimes chicks are not fed by the parents as well as they could be. Perhaps it is a bigger clutch than usual or at worst the hen has died. For whatever reason, chicks, especially the rarer mutations, should not be left to die. Handrearing is a necessary function that should be mastered by all aviculturists.

The following recipe is one that I have used with great success. There are many such recipes available, select one you find successful and stick to it.

MIX A - For Birds up to Pin Feathers

4 oz High protein baby cereal
4 oz Egg and biscuit mix
2 oz Full cream milk powder
1 heaped dessertspoon of equal quantities of maize meal, ground rice, Arrowroot™ biscuit and millet meal (mixed)

For Newborn Chicks - 1 meal daily of plain natural yoghurt mixed with stewed apple (Blue Heinz™) for the first three days.

MIX B - For Older Chicks

8 oz High protein baby cereal
2 oz Sunflower meal
4 oz crushed Arrowroot™ biscuit
4 oz unprocessed wheatgerm
1 teaspoon Glucodin™
1 heaped dessertspoon of equal quantities of maize meal and millet meal (mixed)
Pentavite™ vitamins - 1 drop per bird daily

A little cold water is added and the mixture is stirred into a smooth paste, free of lumps. Boiling water is then added and the mixture stirred to the consistency required - a yoghurt-like consistency is fine. It should run fairly easily off the spoon. To ensure the mix is consistently fine in texture, process in a blender then sieve through a fine sieve.

The amount of handrearing mixture required at each feed (obviously depending on the number of chicks), should be placed in a small dish - a tea cup is ideal. It must be fed to the chicks fairly hot, bearable to your inside wrist or top lip. Refusal to feed often indicates a temperature problem. The small dish or cup is placed in a bowl of hot water to keep the mix warm enough for the last chick. Boiling water is added to the big bowl to heat up the food if required. The amount of food taken may vary from chick to chick. The crop should be filled but not to the extent where it is stretched to the limit. Commonsense is all that is required. If the chick is more than ten days old, the crop should be practically empty prior to each feed.

Four feeds a day should be quite sufficient. The first feed is given between 6am and 8am, the second between 1 pm and 2pm, the third between 5pm and 6pm and the final feed before you go to bed between 10pm and midnight. The chicks will make it through the night without a feed. For chicks less than ten days old, feed when the crops are empty. This could be anywhere between two and five hours apart. Again, even at this age, a feed as late at night as possible and one early in the morning will be satisfactory.

My wife, Jacki, finds the easiest and cleanest method of feeding is with a 2ml dialysis syringe. Place the tip of the syringe into the side of the mouth and gently dispense the food. The chick will swallow it normally. Do not feed directly into the crop. Another successful method is to use a teaspoon bent up at the sides to form a funnel. Wipe the beak, face and throat feathers clean of food after each feed. One of the main reasons our chicks are kept in a warm brooder is that if a bird with warm to hot food in its crop is kept in a cool place, sour crop problems can occur. Since using the brooder, sour crop has never been a problem.

Should you have an outbreak of sour crop, the following steps should be taken:

- Add 3 drops of Mycostatin™ to the food and massage the crop.

- If the crop has seized up completely, use Mycostatin™ in water. Give no food until the crop is clear. Once the crop is clear, feed natural yoghurt and stewed apple with a pinch of Glucodin™ for two to four feeds and then go back to a runny diet with 1 drop of Mycostatin™ added per feed.

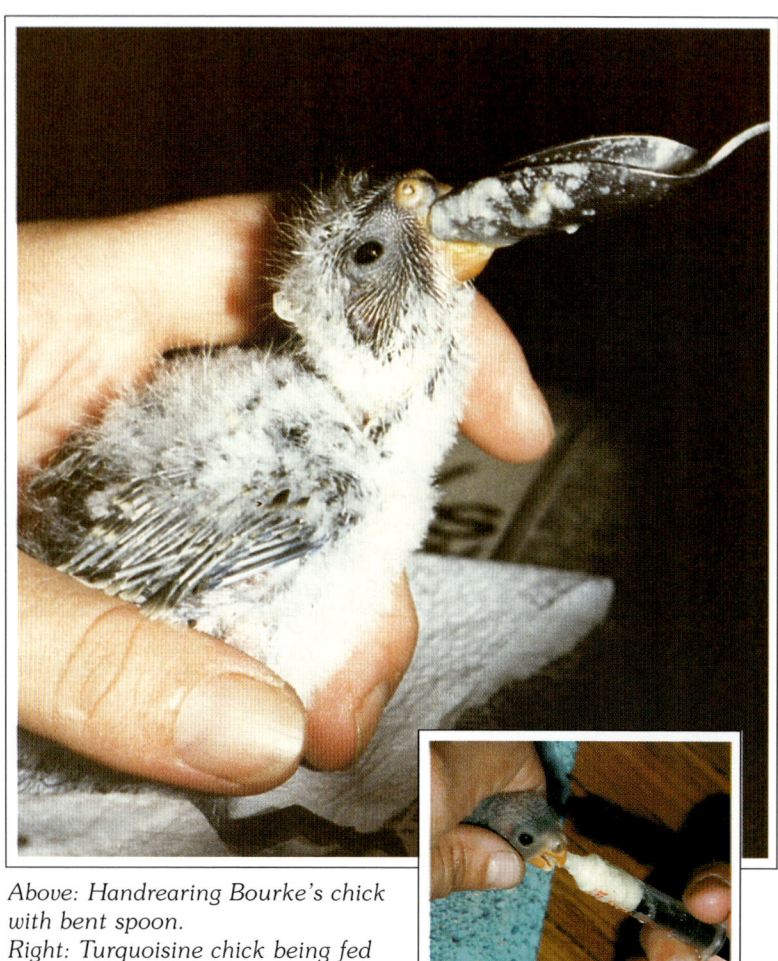

Above: Handrearing Bourke's chick with bent spoon.
Right: Turquoisine chick being fed from a dialysis syringe.

Genetic Tables

It would obviously take up too much space to write out all the possible matings and expectations of the various mutations mentioned in this book. However, the following are some of the most useful matings for the production of secondary mutations.

Pink Bourke's Matings
Combinations of Sex-linked and Recessive mutations.

COCK		HEN
Cream/Rose	x	**Rose/Cream**
1/8 Pink		1/8 Pink
1/8 Cream/Rose		1/8 Cream
1/8 Rose/Cream		1/8 Rose/Cream
1/8 Normal/Rose/Cream		1/8 Normal/Cream
Rose/Cream	x	**Rose/Cream**
1/8 Pink		1/8 Pink
1/4 Rose/Cream		1/4 Rose/Cream
1/8 Rose		1/8 Rose
Rose/Cream	x	**Cream**
1/4 Cream/Rose		1/4 Pink
1/4 Normal/Rose/Cream		1/4 Rose/Cream
Rose/Cream	x	**Pink**
1/4 Pink		1/4 Pink
1/4 Rose/Cream		1/4 Rose/Cream
Normal/Rose/Cream	x	**Pink**
1/8 Pink		1/8 Pink
1/8 Rose/Cream		1/8 Rose/Cream
1/8 Cream/Rose		1/8 Cream
1/8 Normal/Rose/Cream		1/8 Normal/Cream

B. BRANSTON

Pink Bourke's Parrot

Jade Yellow and Olive Yellow Turquoisine Matings

Combinations of Dominant and Recessive mutations. Sex of parents irrelevant, sex of offspring 50:50

Jade/Yellow x **Jade Yellow**
1/8 Olive/Yellow
1/4 Jade/Yellow
1/8 Normal/Yellow

1/8 Olive Yellow
1/4 Jade Yellow
1/8 Yellow

Olive/Yellow x **Jade Yellow**
1/4 Olive/Yellow
1/4 Jade/Yellow

1/4 Olive Yellow
1/4 Jade Yellow

Jade/Yellow x **Olive Yellow**
1/4 Olive/Yellow
1/4 Jade/Yellow

1/4 Olive Yellow
1/4 Jade Yellow

Olive/Yellow x **Yellow**
1/2 Jade/Yellow

1/2 Jade Yellow

Olive Yellow x **Normal/Yellow**
1/2 Jade/Yellow

1/2 Jade Yellow

Right: Olive Yellow Turquoisine Parrot.
Below: Jade Yellow Turquoisine Parrot.

PHOTOGRAPHS BY G. ROMAN

Page 80

Cinnamon Opaline Red-rumped Matings
Combinations of Sex-linked and Sex-linked mutations.

COCK	HEN
Cinnamon x	**Opaline**
1/2 Normal/Cinnamon/Opaline	1/2 Cinnamon
Opaline x	**Cinnamon**
1/2 Normal/Cinnamon/Opaline	1/2 Opaline
Normal/Cinnamon/Opaline x	**Opaline**
1/12 Opaline/Cinnamon	1/12 Cinnamon Opaline
1/6 Opaline	1/6 Cinnamon
1/6 Normal/Cinnamon/Opaline	1/6 Opaline
1/12 Normal/Opaline	1/12 Normal
Normal/Cinnamon Opaline x	**Opaline**
1/6 Opaline/Cinnamon	1/6 Cinnamon Opaline
1/12 Opaline	1/12 Cinnamon
1/12 Normal/Cinnamon/Opaline	1/12 Opaline
1/6 Normal/Opaline	1/6 Normal

These results assume a COV (Cross Over Value) of 33% based on Budgerigar results.

Cinnamon Opaline Red-rumped Parrot.

Page 81

Blue Olive Pied Red-rumped Matings
Combinations of Recessive and Recessive mutations.
Sex of parents irrelevant, sex of offspring even 50:50

Normal/Blue/Pied x **Normal/Blue/Pied**
1/16 Blue Pied   1/8 Pied/Blue   1/16 Pied
1/8 Blue/Pied   1/4 Normal/Blue/Pied 1/8 Normal/Pied
1/16 Blue   1/8 Normal/Blue   1/16 Normal

Normal/Blue/Pied x **Blue/Pied**
1/8 Blue Pied   1/8 Pied/Blue
1/4 Blue/Pied   1/4 Normal/Blue/Pied
1/8 Blue   1/8 Normal/Blue

Normal/Blue/Pied x **Pied/Blue**
1/8 Blue Pied   1/8 Blue/Pied
1/4 Pied/Blue   1/4 Normal/Blue/Pied
1/8 Pied   1/8 Normal/Pied

Blue/Pied x **Pied/Blue**
1/4 Blue Pied   1/4 Pied/Blue
1/4 Blue/Pied   1/4 Normal/Pied/Blue

Normal/Blue/Pied x **Blue Pied**
1/4 Blue Pied   1/4 Pied/Blue
1/4 Blue/Pied   1/4 Normal/Blue/Pied

Blue/Pied x **Blue Pied**
1/2 Blue Pied   1/2 Blue/Pied

Pied/Blue x **Blue Pied**
1/2 Blue Pied   1/2 Pied/Blue

*Blue Pied Red-rumped hen and
UK Pied Red-rumped Parrots.*

Cinnamon Blue Red-rumped matings
Combinations of Sex-linked and Recessive mutations.

COCK   **HEN**

Blue/Cinnamon   x   **Cinnamon/Blue**
1/8 Cinnamon Blue   1/8 Cinnamon Blue
1/8 Blue/Cinnamon   1/8 Blue
1/8 Cinnamon/Blue   1/8 Cinnamon/Blue
1/8 Normal/Cinnamon/Blue   1/8 Normal/Blue

Cinnamon/Blue   x   **Cinnamon/Blue**
1/8 Cinnamon Blue   1/8 Cinnamon Blue
1/4 Cinnamon/Blue   1/4 Cinnamon/Blue
1/8 Cinnamon   1/8 Cinnamon

Cinnamon/Blue   x   **Blue**
1/4 Blue/Cinnamon   1/4 Cinnamon Blue
1/4 Normal/Blue/Cinnamon   1/4 Cinnamon/Blue

Cinnamon/Blue   x   **Cinnamon Blue**
1/4 Cinnamon Blue   1/4 Cinnamon Blue
1/4 Cinnamon/Blue   1/4 Cinnamon/Blue

Normal/Cinnamon/Blue   x   **Cinnamon Blue**
1/8 Cinnamon Blue   1/8 Cinnamon Blue
1/8 Cinnamon/Blue   1/8 Cinnamon/Blue
1/8 Blue/Cinnamon   1/8 Blue
1/8 Normal/Cinnamon/Blue   1/8 Normal/Blue

Cinnamon Blue Red-rumped hen and cock.

Cinnamon Blue Opaline Red-rumped Matings

Combinations of Sex-linked, Recessive and Sex-linked mutations.

It is best to start with:

COCK		HEN
Cinnamon Opaline	x	**Blue**
1/2 Normal/Cinnamon Opaline/Blue		1/2 Cinnamon Opaline/Blue

then breed

Normal/Cinnamon Opaline/Blue	x	**Cinnamon Opaline/Blue**
1/24 Cinnamon Opaline Blue		1/24 Cinnamon Opaline Blue
1/12 Cinnamon Opaline/Blue		1/12 Cinnamon Opaline/Blue
1/24 Cinnamon Opaline		1/24 Cinnamon Opaline
1/48 Cinnamon Blue/Opaline		1/48 Cinnamon Blue
1/24 Cinnamon/Opaline/Blue		1/24 Cinnamon/Blue
1/48 Cinnamon/Opaline		1/48 Cinnamon
1/48 Opaline Blue/Cinnamon		1/48 Opaline Blue
1/24 Opaline/Cinnamon/Blue		1/24 Opaline/Blue
1/48 Opaline/Cinnamon		1/48 Opaline
1/24 Blue/Cinnamon Opaline		1/24 Blue
1/12 Normal/Cinnamon Opaline/Blue		1/12 Normal/Blue
1/24 Normal/Cinnamon Opaline		1/24 Normal

The alternative starts of Cinnamon Blue x Opaline or Opaline Blue x Cinnamon would take approximately three generations to reliably produce Cinnamon Opaline/Blue hens. Once that is achieved the following generation will produce 1/48 Cinnamon Opaline Blue cocks.

Normal/Cinnamon/ Opaline/Blue	x	**Cinnamon Opaline/Blue**
1/48 Cinnamon Opaline Blue		1/48 Cinnamon Opaline Blue
1/24 Cinnamon Opaline/Blue		1/24 Cinnamon Opaline/Blue
1/48 Cinnamon Opaline		1/48 Cinnamon Opaline
1/24 Cinnamon Blue/Opaline		1/24 Cinnamon Blue
1/12 Cinnamon/Opaline/Blue		1/12 Cinnamon/Blue
1/24 Cinnamon/Opaline		1/24 Cinnamon
1/24 Opaline Blue/Cinnamon		1/24 Opaline Blue
1/12 Opaline/Cinnamon/Blue		1/12 Opaline/Blue
1/24 Opaline/Cinnamon		1/24 Opaline
1/48 Blue/Cinnamon Opaline		1/48 Blue
1/24 Normal/Cinnamon Opaline/Blue		1/24 Normal/Blue
1/48 Normal/Cinnamon Opaline		1/48 Normal

A Word of Warning

There is no doubt that there will be increases in the number of mutations in the birds discussed in this book. Established colours will become cheaper and more widely available. Let us make sure that we do not lose our lines of pure Normal birds. Not only must the original birds be maintained for their own sake but also for the fact that many mutations, especially newer ones, must be bred back through good Normals. This ensures that mutations remain healthy, vigorous and fertile and also conform to the standard of the Normal bird.

Let us hope we have learned something in this regard from the African Lovebird mutations and not repeat any of the mistakes of mass producing inferior birds.

We can begin now by rectifying our first mistake which reduced the Yellow Turquoisine to a weedy little bird. This was a prime example of what harm the 'dollar' can do to birds in certain peoples' hands.

Many breeders are now turning to Grass Parrots and their mutations. If some of these breeders go for quantity rather than quality then buyers should be adamant and not purchase inferior birds. The breeders who worship the dollar rather than the bird will only change their ways if the market dictates it.

Cinamon Blue Opaline Red-rumped cock.

The Acclaimed 'A Guide to ...' range

Concise, informative and colourful reading for all bird keepers and aviculturists.

- A Guide to Gouldian Finches
- A Guide to Australian Long and Broad-tailed Parrots and New Zealand Kakarikis
- A Guide to Rosellas and Their Mutations
- A Guide to Eclectus Parrots
- A Guide to Cockatiels and Their Mutations
- A Guide to Pigeons, Doves and Quail
- A Guide to Neophema and Psephotus Grass Parrots and Their Mutations (Revised Edition)
- A Guide to Asiatic Parrots and Their Mutations (Revised Edition)
- A Guide to Zebra Finches
- A Guide to Australia Grassfinches
- A Guide to Basic Health and Disease in Birds (Revised Edition)
- A Guide to Incubation and Handraising Parrots
- A Guide to Pheasants and Waterfowl
- A Guide to Pet and Companion Birds
- A Guide to Australian White Cockatoos
- A Guide to Popular Conures
- A Guide to Lories and Lorikeets (Revised Edition)
- A Guide to Colour Mutations and Genetics in Parrots
- A Guide to Macaws as Pet and Aviary Birds

Simply the best publications on pet & aviary birds available ...

Australian BirdKeeper MAGAZINE

Six glossy, colourful and informative issues per year. Featuring articles written by top breeders, bird trainers, avian psychologists and avian veterinarians from all over the world.

SUBSCRIPTIONS AVAILABLE

For subscription rates and FREE catalogue contact ABK Publications.

Handbook of Birds, Cages & Aviaries

One of the most popular cage/aviary bird publications ever produced.

ABK Publications stock a complete and ever increasing range of books and videos on all pet and aviary birds.

For further information or Free Catalogue contact:

ABK PUBLICATIONS

P.O. Box 6288
South Tweed Heads
NSW 2486 Australia
Phone: (07) 5590 7777 Fax: (07) 5590 7130
Email: birdkeeper@birdkeeper.com.au
Website: www.birdkeeper.com.au

Publishers Note

This title is published by **ABK Publications** who produce a wide and varied range of avicultural literature including the world acclaimed **Australian Birdkeeper** magazine - a full colour, bi-monthly magazine specifically designed for birdlovers and aviculturists. It is the intention of the publishers to produce high quality, informative literature for birdlovers, fanciers and aviculturists alike throughout the world. It is also the publishers' belief that the dissemination of qualified information on the care, keeping and breeding of birds is imperative for the total well-being of captive birds and the increased knowledge of aviculturists.

Nigel Steele-Boyce
Publisher/Editor-In-Chief
ABK Publications

For further information or Free Catalogue contact:

ABK Publications
P.O. Box 6288
South Tweed Heads
NSW 2486 Australia

Phone: (07) 5590 7777 Fax: (07) 5590 7130
Email: birdkeeper@birdkeeper.com.au
Website: www.birdkeeper.com.au